# FRACTURED AQUIFERS

## Formation evaluation by well testing

Kurt Ambo Nielsen

The cover photo by Christian Knudsen shows ground water flowing from fractured limestone on the island of Amager near Copenhagen. From *Zinck-Jorgensen, K. and K. Hinsby, 2001: Fractures, water and oil in limestone. News from the Geological Survey of Denmark and Greenland No. 1, p. 10-13 (In Danish).*

Visit the author's website at www.kanmiljoe.dk

Order this book online at www.trafford.com/07-1087
or email orders@trafford.com

Most Trafford titles are also available at major online book retailers.

© Copyright 2007 Kanmiljoe.
All rights reserved. No part of this publication may be reproduced, stored in a retrieval system, or transmitted, in any form or by any means, electronic, mechanical, photocopying, recording, or otherwise, without the written prior permission of the author.

Note for Librarians: A cataloguing record for this book is available from Library and Archives Canada at www.collectionscanada.ca/amicus/index-e.html

Printed in Victoria, BC, Canada.

ISBN: 978-1-4251-3019-0

*We at Trafford believe that it is the responsibility of us all, as both individuals and corporations, to make choices that are environmentally and socially sound. You, in turn, are supporting this responsible conduct each time you purchase a Trafford book, or make use of our publishing services. To find out how you are helping, please visit www.trafford.com/responsiblepublishing.html*

*Our mission is to efficiently provide the world's finest, most comprehensive book publishing service, enabling every author to experience success. To find out how to publish your book, your way, and have it available worldwide, visit us online at www.trafford.com/10510*

**Trafford**
PUBLISHING   www.trafford.com

**North America & international**
toll-free: 1 888 232 4444 (USA & Canada)
phone: 250 383 6864 ♦ fax: 250 383 6804 ♦ email: info@trafford.com

**The United Kingdom & Europe**
phone: +44 (0)1865 722 113 ♦ local rate: 0845 230 9601
facsimile: +44 (0)1865 722 868 ♦ email: info.uk@trafford.com

10 9 8 7 6 5 4 3 2

# Preface

Ask an engineer, a geologist or any person with a scientific degree to work with problems he or she does not understand, and he or she will be annoyed. I have been annoyed many times when it was my task to interpret pumping test data from wells tapping fractured aquifers. Data simply did not behave as they should according to the textbooks. More often than not, I had to apply the standard excuse as if that explains anything: We are dealing with geological inhomogeneities.

When I finally found the courage and time to consult the oil literature, things started to fall into place. Actually, I have not yet come across any test data from fractured wells, which did not fit one of the models developed by oil engineers.

Oil engineers are pragmatic people. They look at a complicated fracture system and reduce it to a homogeneous system of blocks or layers. Then they set up a mathematical model which, matched with the data, gives them information about fractures and matrix. That gives them numbers to work with.

Some researchers claim that such simplifications are not justified. There are no straight lines in nature. In addition, sometimes more than one model fits data and renders the interpretation ambiguous. I am of the old school. I prefer to have numbers to work with rather than nothing at all.

The models work, however, and it is then up to the reader to decide for himself whether or not he believes in the results.

The following Danish private and municipal water supplies and authorities have kindly made valuable field data available for this book: Copenhagen, Odense, Koege, Hoeje Taastrup, Greve Strand, Region Bornholm, Region Roskilde, Esbjerg , Roedovre and the County of Copenhagen. Without this data, this book would not exist.

Gyrite Brandt, Copenhagen Energy, Denmark and Daniel Lufkin, Maryland, USA have proof read the manuscript. I owe my gratitude to both for valuable comments and suggestions.

Copenhagen, 2007


Kurt Ambo Nielsen

# Contents

1. Introduction ..................................................................................................... 7
2. The homogeneous aquifer ............................................................................... 9
3. Flow in fractures ............................................................................................ 13
   3.1 Ideal fractures ........................................................................................... 13
   3.2 Natural fractures ....................................................................................... 14
   3.3 Fracture systems ....................................................................................... 14
4. Skin effect and well performance .................................................................. 17
   4.1 Skin effect ................................................................................................. 17
   4.2 Well performance ..................................................................................... 21
   4.3 Partial penetration .................................................................................... 23
   4.4 Skin calculations from recovery data ....................................................... 25
   4.5 Well bore storage ...................................................................................... 29
5. Pressure-dependent transmissivity ................................................................ 31
   5.1 Sorensen's model ...................................................................................... 31
     5.1.1 Drawdown data .................................................................................. 31
     5.1.2 Recovery data .................................................................................... 33
     5.1.3 Discussion .......................................................................................... 33
6. The dual-porosity aquifer ............................................................................... 35
   6.1 Warren and Root ....................................................................................... 36
     6.1.1 The significance of $\lambda$ and $\omega$ .................................................................... 41
   6.2 de Swaan .................................................................................................. 41
   6.3 Kazemi ...................................................................................................... 45
   6.4 Bourdet and Gringarten ............................................................................ 48
   6.5 Streltsova ................................................................................................... 52
   6.6 Najurieta ................................................................................................... 62
     6.6.1 The stratum case ................................................................................ 62
     6.6.2 Evaluation of hydraulic parameters for the pumped well ................. 68
     6.6.3 Drawdown in the aquifer ................................................................... 74
     6.6.4 The blocks case .................................................................................. 78
   6.7 Estimating block size or layer thickness .................................................. 83
7. Single vertical fracture .................................................................................. 85
   7.1 The physical model .................................................................................. 86
   7.2 The uniform flux vertical fracture ............................................................ 87
     7.2.1 Drawdown in the pumped well ......................................................... 87
     7.2.2 Drawdown in the aquifer ................................................................... 89
     7.2.3 Evaluation of hydraulic parameters ................................................... 93
   7.3 The vertical fracture with infinite hydraulic conductivity ....................... 94

    7.4 Effective radius and skin factor .................................................................. 96
    7.5 Vertical fracture with finite hydraulic conductivity ........................................ 100
        7.5.1 Effective radius and skin factor ...................................................... 104
        7.5.2 Radius of influence for a vertical fracture ...................................... 105
    7.6 Partially penetrating vertical fractures ........................................................ 109
    7.7 A well near a vertical fracture .................................................................... 110
    7.8 A well in a vertical dike ............................................................................. 120
    7.9 Jenkins and Prentice ................................................................................. 121
8. Single horizontal fracture ................................................................................. 129
    8.1 The physical model ................................................................................... 129
    8.2 Evaluation of hydraulic parameters ........................................................... 133
    8.3 Pseudo-skin factor, skin factor and effective radius .................................. 134
9. Diagnostics ...................................................................................................... 141
10. Brief guidelines to fractured aquifer testing .................................................... 147
11. Variable discharge ......................................................................................... 149
    11.1 The continuous line source ...................................................................... 150
    11.2 Exponentially varying discharge .............................................................. 152
        11.2.1 Exponential decrease .................................................................. 152
        11.2.2 Exponential increase ................................................................... 156
    11.3 Hyperbolically varying discharge ............................................................. 159
        11.3.1 Hyperbolic decrease .................................................................... 159
        11.3.2 Hyperbolic increase ..................................................................... 162
    11.4 Linear change in discharge ..................................................................... 162
11.5 A discharge curve with a maximum value ..................................................... 167
Literature ............................................................................................................. 173
List of symbols .................................................................................................... 179
Index .................................................................................................................... 183
Annex 1. Problems .............................................................................................. 185
Annex 2. Type-curve tables ................................................................................. 211

# 1. Introduction

Ground water flow in fractured aquifers behaves very differently from ground water flow in aquifers where the porous medium consists of sand. In sand aquifers, there is one single flow regime, in fractured aquifers there are two, fractures and matrix. This difference is very important to appreciate when interpreting data from well tests in fractured aquifers. When we stress a fractured aquifer by pumping, early drawdown data will show the fracture character and late data will show the combined fracture and matrix system behavior of the aquifer. The transition zone connecting early and late aquifer response reveals matrix properties and needs careful analyses to evaluate the aquifer parameters. The benefit of a correct data interpretation is valuable information about fracture as well as matrix properties. This matrix-fracture behavior is fundamental in the oil industry. Therefore, they developed most of the methods available for data interpretation. The oil industry is particularly concerned with whether the oil occurs in fractures or in matrix, and the methods developed by oil engineers focus on this relationship.

In the 1960s the first groundbreaking research on fractured reservoirs appeared and dealt with uniformly fractured dual-porosity reservoirs. Much of this research is fundamental and is still in use. Sometimes a single horizontal or vertical fracture controls the ground water flow to wells. Methods of analyses of pumping test data from wells tapping such fractures were published in the 1970's. The ground water industry apparently took some time to adopt the methods and the first research on fractured aquifers appears in the 1980s.

Unfortunately, the oil field notation, units and parameters are much different from those used in the ground water sector, and therefore ground water professionals tend not to consult the oil literature, the result being that the interpretation methods are rarely used in the ground water sector. This book bridges the gap between the two sectors and presents the state-of-the-art methods for interpreting well test data from fractured aquifers using ground water units and notation. The book outlines the essential mathematics behind the methods, and numerous field cases demonstrate the interpretation methods. Many pitfalls exist when interpreting data and, in some cases, a unique interpretation is not possible. The book explains the behavior of pumped wells and monitoring wells for several different types of uniformly fractured aquifers, aquifers with a single horizontal or vertical fracture, and aquifers with parameters changing with pressure. Methods for evaluation of the efficiency of wells and a basic diagnostics tool are included.

The book includes a problems section based on real data, which may be difficult to come by otherwise. The problems are for use in classroom sessions or as training for the practicing engineer.

The last chapter gives analytical methods for analyzing historical data from well fields. More often than not, the discharge rate has varied over time rather than being constant. This requires preparation of type-curves with time-dependent discharge rate. Several cases and examples illustrate the method.

## 2. The homogeneous aquifer

When a well taps a fractured aquifer, the early drawdown data exhibits the fracture properties. After a short time, ground water flows from matrix into fractures and reduces the rate of drawdown. After sufficient time the radius of the cone of depression will become large compared to the characteristic length of the fracture system and the drawdown now behaves as if there is only one flow regime, namely the combined fracture and matrix system. In this book we refer to this late period of the well test as homogeneous aquifer behavior. (*Gringarten, 1984*) related the term homogeneous to aquifer behavior, not to aquifer geology. Homogeneous in his terminology means that the hydraulic conductivity in a test and that measured in a core are the same. When the aquifer behavior is homogenous, the flow to the well is radial or pseudo-radial and described mathematically by the classical Theis solution (*Theis, 1935*),

$$s = \frac{Q}{4\pi T} \int_u^\infty \frac{e^{-x}}{x} dx, \qquad u = \frac{r^2 S}{4Tt} \qquad (2.1)$$

where

       $s$ is the drawdown in the pumped well or in a monitoring well (m)
       $Q$ is the discharge rate (m³/s)
       $T$ is the aquifer transmissivity (m²/s)
       $S$ is the aquifer storage coefficient (dimensionless)
       $r$ is the well radius or distance to monitoring wells (m)
       $t$ is the time since pumping started (s)

Here, the usual assumptions apply: the aquifer is homogeneous and isotropic, unlimited in areal extent, the well utilizes the full aquifer thickness and the water is released instantaneously from aquifer storage due to elastic compression (no well bore storage).

The integral is known as the exponential integral, and the mathematical notation is $-Ei(-u)$. In the oil industry, the Theis solution is known as *the exponential integral solution* or *the line source solution*. In ground water, the exponential integral is known as the well function $W(u)$. *Cooper and Jacob (1946)* showed that the series representation of the integral for $u < 0.01$ (the Jacob – Cooper equation) is:

$$s = \frac{Q}{4\pi T}(-0.5772 - \ln u) = \frac{Q}{4\pi T}\ln\frac{0.561}{u} = \frac{2.3Q}{4\pi T}\log\frac{2.25Tt}{r^2 S} = \frac{0.183Q}{T}\log\frac{2.25Tt}{r^2 S} \qquad (2.2)$$

It is important that Equation 2.2 be used only when $u < 0.01$, that is for small values of $r^2/t$. If this condition is not met, gross errors may be made when calculating hydraulic parameters. Equation 2.2 is always valid for the pumped well, but its validity should be checked when applied to monitoring well data.

In Equation 2.2, units are consistent (meters and seconds). When plotting pumping test data time is usually in minutes and Equation 2.2 for practical applications becomes

$$s = \frac{0.183Q}{T}\log\frac{135Tt}{r^2 S} \qquad (2.3)$$

Here, $Q$ is in m³/s, $t$ in minutes, $r$ in meters, and the result is $T$ in m²/s and $S$ is dimensionless. Equation 2.3 shows the well-known log-log relationship between drawdown and time. A plot of drawdown against $\log t$ produces a straight line with slope $\Delta s$ per log cycle. Based on this, we can use a graphical method to calculate the transmissivity. For monitoring wells, we use the intersection $t_0$ with the $s = 0$ line to calculate the storage coefficient, see Figure 2.1.

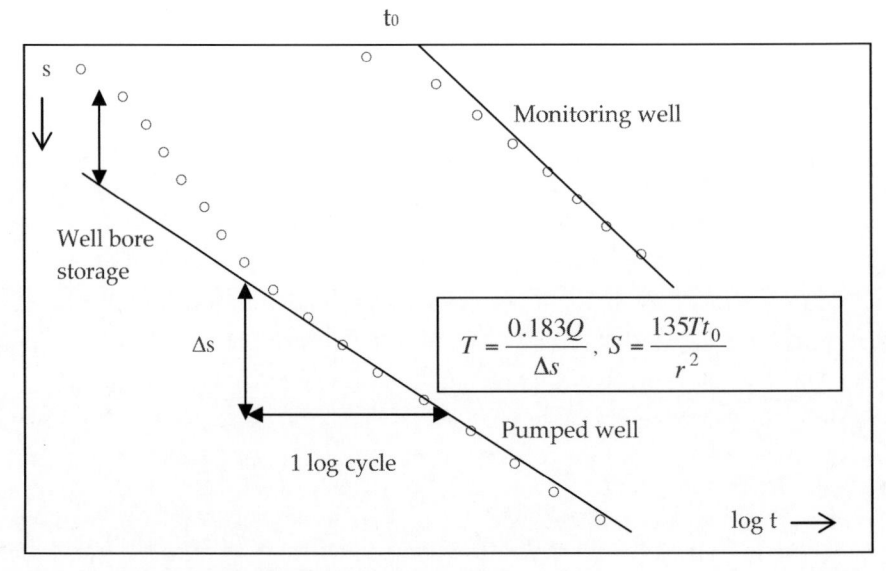

*Figure 2.1 Calculation of T and S*

We recognize these equations from the classical well test analyses, but they also play an important role in formation evaluation of fractured aquifers. The significance of the physical parameters $T$ and $S$ vary with the type of fracture system characterizing the aquifer, and they are used to describe fractures as well as matrix. Indeed, the whole purpose of well test analysis in fractured aquifers is to achieve a physical description of the system and hereby establish the means to understand the aquifer behavior and prepare correct forecasts of the future drawdowns in the aquifer.

In ground water hydraulics the following dimensionless parameters are traditionally used:

Dimensionless drawdown: $s_D = W(u) = \dfrac{4\pi T s}{Q}$ \hfill (2.4)

Dimensionless time: $t_D = 1/u = \dfrac{4Tt}{r^2 S}$ \hfill (2.5)

In the oil industry, the following parameters are traditionally used:

Dimensionless drawdown: $s_D = \dfrac{2\pi T s}{Q}$ \hfill (2.6)

Dimensionless time: $t_D = \dfrac{Tt}{r^2 S}$ \hfill (2.7)

In oil industry notation, Equation (2.2) therefore, is:

$$s_D = \frac{1}{2}\left(\ln(4t_D) - 0.5772\right) = \frac{1}{2}\left(\ln t_D + 0.80907\right), \quad t_D > 25 \qquad (2.8)$$

In principle it is irrelevant which dimensionless parameters we use, but in this manual we always use the ground water notation in Equation 2.4 for dimensionless drawdown, and where possible Equation 2.5 for dimensionless time. The latter, however, will not always be the case when producing type-curves because the tabular values presented in the literature are often based on $t_D$ in Equation 2.7.

# 3. Flow in fractures
## 3.1 Ideal fractures

Characteristically, fractured aquifers have a high fracture hydraulic conductivity and a low matrix hydraulic conductivity as opposed to high matrix storage and low fracture storage. For obvious reasons there seem to be no full-scale field studies of fracture geometry and hydraulics. Available knowledge about flow in fractures is based on laboratory experiments with artificial fracture systems and in some cases with field samples. The following overview of the theory of fracture flow is based on the work presented by *Sharp and Maini (1972)*, who include an extensive list of literature on the subject.

The theory of flow in ideal fractures assumes fractures with known geometry namely, plane fractures with known aperture. It is generally assumed that the relationship between the flow rate $q$ and the hydraulic gradient $I$ in the fracture is

$$q = K_f \cdot I, \qquad (3.1)$$

which is analogous to Darcy's law for laminar flow in porous media. The factor of proportionality $K_f$ is the fracture hydraulic conductivity.

The exact solution to the laminar viscous flow of a fluid between two parallel plates is found by solving Navier-Stoke's equations:

$$q = \frac{\rho g e^3}{12 \mu} I, \qquad (3.2)$$

where $q$ is the fluid velocity, $e$ is the fracture aperture, $\mu$ is the dynamic viscosity, $\rho$ is the density and $g$ the acceleration of gravity. This cubic law shows that the flow rate increases with the third power of the aperture, which is the reason why fractured wells may have very high yields. Especially karstic limestone may have very high transmissivities up to 0.1 m²/s. On the other hand, limestone with low hydraulic conductivity fractures may have transmissivity values below 10⁻⁴ m²/s.

Reynolds number $R$ for fracture flow is analogous to Reynolds number for flow in inter-granular media with the grain size diameter replaced by the fracture aperture:

$$R = \frac{q \rho g e}{\mu} \qquad (3.3)$$

The literature describes attempts to find representative values of the aperture *e* in fractures with rough walls. *Scheidegger (1960)* compared the results from a number of studies and found them to be inadequate. The use of Reynolds number to describe the nature of flow is therefore restricted to fractures with a roughness that is small compared to the aperture, and also the flow must be parallel and the fracture cannot have a curvature. Experiments performed by *Louis (1969)*, however, showed that in parallel flow, turbulent flow results when $R > 2300$, as opposed to inter-granular flow, where turbulence takes place for $R > 1\text{-}10$.

## 3.2 Natural fractures

In natural fractures, the aperture is not constant, the roughness may be large compared to the aperture and the flow pattern can be far from parallel. In places, the flow may be three-dimensional. This means that the cubic law, Equation 3.2, may not rule. The flow regime in the fracture may change from laminar to turbulent and back resulting in a non-linear relationship between $q$ and $I$.

*Sharp (1970)* carried out a number of experiments with different fracture geometries and concluded that the relationship between flow rate and gradient cannot be described by a universal law, and that the flow through fractures with large roughness does not resemble the flow between parallel plates with moderate roughness. Sharp concluded that for apertures in the range 0.25-1.5 mm the values of $n$ in the formula $q \propto e^n$ are the following,

|  | Rough fracture | Smooth parallel fracture |
|---|---|---|
| Linear laminar flow | $n \sim 2$ | $n = 3$ |
| Non-linear laminar flow | $1.2 < n < 2$ | - |
| Fully turbulent flow | $n \sim 1.2$ | $n = 1.5$ |

Thus, it appears that the flow rate in the fracture increases faster than the fracture aperture.

## 3.3 Fracture systems

Natural fractures form an irregular pattern of interconnected or isolated conduits of ground water. Normally, we can assume that over a large aquifer volume the fractures constitute a system with a fracture volume that is small compared to the bulk volume. If we assume that Darcy's law is valid for individual fractures, it is also valid for the fracture system. Taken over a sufficiently large volume it appears appropriate to consider the flow similar to that in inter-granular porous media. This leads to the concept of the homogeneous aquifer mentioned earlier, where the matrix block dimensions are small compared to the radius of the cone of depression. When calculating hydraulic parameters we shall take the presence of fractures into consideration, even though the drawdown curves may be similar to those obtained from sand aquifers. The homogeneous aquifer will reveal itself after a sufficiently long time of pumping. For short times, fracture flow dominates and it is possible by applying correct methods of interpretation to evaluate the nature and orientation of fractures. To complete the aquifer evaluation it is necessary to obtain drawdown data from monitoring wells, and with such data at hand, it will be possible to give a description of fracture and matrix hydraulic parameters.

# 4. Skin effect and well performance
## 4.1 Skin effect

*Van Everdingen (1953)* and *Hurst (1953)* introduced the skin effect concept as a possible explanation of why, in abstraction wells, analytical solutions given by Equation 2.8 often match field data poorly. The skin effect is a result of drilling mud invading the formation during drilling creating a low hydraulic conductivity zone around the well. This zone may form a skin or a zone with reduced hydraulic conductivity.

The skin loss $s_s$ is a well loss and is widely used in the analyses of production wells in the oil industry. The skin loss is a linear loss i.e. proportional to $Q$. Water supply wells often have high abstraction rates and therefore, turbulent losses, i.e. losses proportional to $Q^2$, are common. In spite of that, skin effect calculations in fractured water supply wells are very useful to obtain an understanding of the influence of fractures on well performance. Often, well losses in fractured wells are very small and for such wells, skin calculations provide useful results.

Equation 2.8 defines the skin effect through a skin factor $\sigma$ defined by:

$$\frac{2\pi T s_w}{Q} = \frac{1}{2}\left(\ln t_D + 0.80907\right) + \sigma, \qquad (4.1)$$

where $s_w$ is the drawdown in the pumped well. The skin loss is therefore

$$s_s = \sigma \frac{Q}{2\pi T} \qquad (4.2)$$

The drawdown in the pumped well now reads

$$s_w = \frac{Q}{4\pi T}\left(\ln t_D + 0.80907 + 2\sigma\right) \qquad (4.3)$$

In Equation 4.3, the screen radius $r_w$ is used as distance. There is a certain distance from the well where the skin effect vanishes, i.e. $s_s = 0$. This distance represents the effective well radius $r_{ef}$. The effective radius is calculated from

$$r_{ef} = r_w e^{-\sigma} \qquad (4.5)$$

This is easily seen by inserting $r_w = r_{ef} e^\sigma$ in Equation 4.3,

$$s_w = \frac{Q}{4\pi T}\left(\ln\frac{Tt}{Sr_{ef}^2 e^{2\sigma}} + 0.80907 + 2\sigma\right)$$

$$= \frac{Q}{4\pi T}\left(\ln\frac{Tt}{r_{ef}^2 S} + 0.80907 + 2\sigma - 2\sigma\right)$$

For positive values of $\sigma$ we get $r_{ef} < r_w$ and the well is considered under-stimulated and can be improved by development. If $\sigma < 0$ we get $r_{ef} > r_w$, and the well is stimulated, see Figure 4.1.

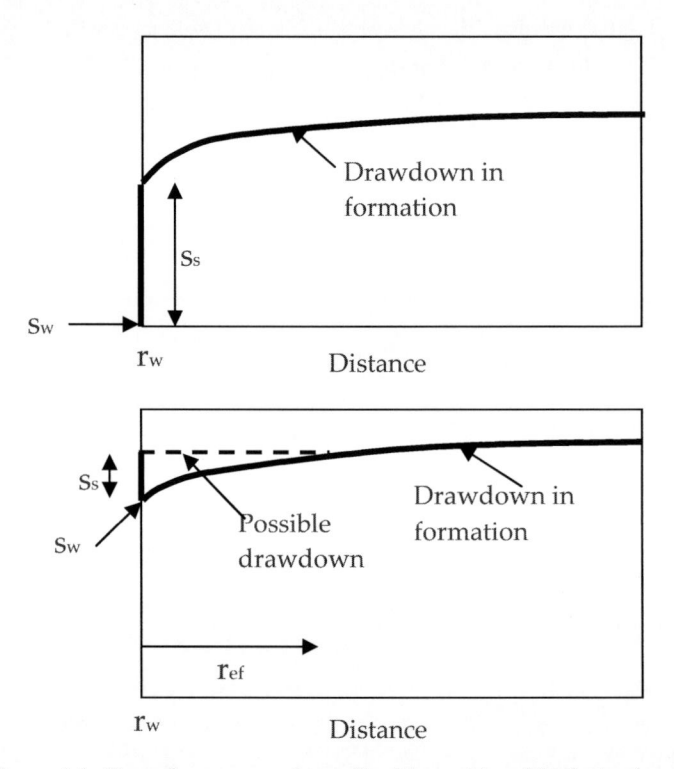

*Figure 4.1. Drawdown around a well with positive skin factor (under-stimulated well, top), and around a well with negative skin factor (stimulated well, bottom). After Earlougher, (1977).*

If the mud invasion has not resulted in a skin but a finite zone with thickness $r_s$ with hydraulic conductivity $K_s$, the skin factor becomes (Hawkins, 1956),

$$\sigma = \left(\frac{K}{K_s} - 1\right)\ln\frac{r_s}{r_w} \qquad (4.6)$$

Either $\sigma$, $K_s$ or $r_s$ may be determined if the two other parameters are known. The skin loss calculation from Equation 4.2 still applies.

Applying Equation 2.2 and using ground water notation, Equation 4.3 reads

$$s_w = \frac{0.183Q}{T}\left(\log\frac{135Tt}{r_w^2 S} + 0.8686\sigma\right) \qquad (4.7)$$

The skin factor becomes

$$\begin{aligned}
\sigma &= 1.1513\left(\frac{s_w T}{0.183Q} - \log\frac{135Tt}{r_w^2 S}\right) = \\
&\quad 1.1513\left(\frac{s_w T}{0.183Q} - \log t - \log\frac{135T}{r_w^2 S}\right) = \\
&\quad 1.1513\left(\frac{s_w}{\Delta s_w} - \log t - \log\frac{135T}{r_w^2 S}\right) = \\
&\quad 1.1513\left(\frac{s_w}{\Delta s_w} - \log t - \log\frac{T}{r_w^2 S} - 2.13\right)
\end{aligned} \qquad (4.8)$$

where $\Delta s_w = \dfrac{0.183Q}{T}$ is the log cycle slope in a semi-log plot of drawdown as a function of time. From Equation 4.8 the skin factor $\sigma$ is calculated by selecting an arbitrary value of time $t$ in minutes and use the corresponding well drawdown $s_w$, after first having calculated $\Delta s_w$ and consequently $T$ from the plot. Analysis of monitoring well data provides a calculation of the storage coefficient $S$. In the absence of such data, we estimate the value from experience. It is imperative that the point used to select the time value falls on the straight line used for determination of the transmissivity, because the skin factor calculation requires that there be a log-log relationship between drawdown and time (homogeneous aquifer conditions).

FRACTURED AQUIFERS

Example 4.1 Calculation of skin factor

The example is from a test on a sand stone well on the Island of Bornholm, Denmark *(Nielsen, 2005 a)*. The following data is given: $Q = 35$ m³/hr, from monitoring well data, $S = 0.00112$, $r_w = 0.1225$ m.

| $t$ (min) | $s_w$ (m) | $t$ (min) | $s_w$ (m) | $t$ (min) | $s_w$ (m) | $t$ (min) | $s_w$ (m) |
|---|---|---|---|---|---|---|---|
| 10 | 11.293 | 50.25 | 11.562 | 300 | 12.016 | 2100 | 12.876 |
| 12 | 11.322 | 60.25 | 11.596 | 360 | 12.102 | 2880 | 13.156 |
| 14 | 11.37 | 75.25 | 11.64 | 420 | 12.096 | 3600 | 13.186 |
| 16 | 11.388 | 90 | 11.716 | 540 | 12.162 | 4320 | 13.276 |
| 18 | 11.386 | 105 | 11.718 | 720 | 12.346 | 5,760 | 13.486 |
| 20.25 | 11.416 | 120 | 11.736 | 900 | 12.526 | 7200 | 13.5 |
| 25.25 | 11.466 | 150 | 11.806 | 1140 | 12.642 | 10090 | 13.71 |
| 30.25 | 11.477 | 180 | 11.85 | 1440 | 12.789 | 12970 | 13.752 |
| 40.25 | 11.526 | 240 | 11.916 | 1740 | 12.876 | 17290 | 13.846 |

The log cycle slope is 1.17 m and the transmissivity becomes

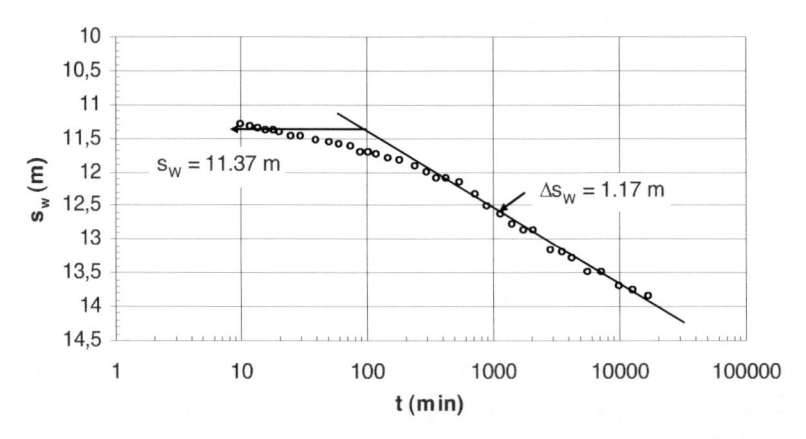

*Figure 4.2. Drawdown data from a fractured sandstone well.*

$$T = \frac{0.183Q}{\Delta s_w} = \frac{0.183 \cdot 35}{1.17 \cdot 3600} = 0.00152 \text{ m}^2/\text{s}.$$

For $t = 100$ min, the drawdown is $s_w = 11.37$ m, and the skin factor is calculated from Equation 4.8:

$$\sigma = 1.1513\left(\frac{s_w}{\Delta s_w} - \log t - \log \frac{T}{r_w^2 S} - 2.13\right) =$$

$$1.1513\left(\frac{11.37}{1.17} - \log 100 - \log \frac{0.00152}{0.1225^2 \cdot 0.00112} - 2.13\right) = 4.18$$

From Equation 4.2 the skin loss becomes

$$s_s = \sigma \frac{Q}{2\pi T} = 4.18 \frac{35}{2\pi \cdot 0.00152 \cdot 3600} = 4.26 \text{ m.}$$

The effective radius is

$$r_{ef} = r_w e^{-\sigma} = 0.1225 e^{-4.26} = 0.0017 \text{ m.}$$

This well is clearly under-stimulated.

## 4.2 Well performance

In the classical well-test literature, the drawdown in the abstraction well is the sum of a linear formation loss and a turbulent loss

$$s_w = BQ + CQ^2 \qquad (4.9)$$

where $B$ is the formation loss factor and $C$ is the well loss factor. From Equation 4.9 the well efficiency becomes

$$V = \frac{BQ}{BQ + CQ^2} \qquad (4.10)$$

In the oil literature, the well efficiency is based on the skin loss:

$$V = \frac{s_w - s_s}{s_w} \qquad (4.11)$$

Oil reservoir engineers also define a damage ratio or a clogging index $C_i$, which is the reciprocal of the well efficiency:

$$C_i = 1/V = \frac{s_w}{s_w - s_s}, \qquad (4.12)$$

and finally a damage factor or clogging factor $C_f$, derived from the well efficiency:

$$C_f = 1 - V = \frac{s_s}{s_w} \qquad (4.13)$$

As mentioned before, the use of these oil literature well performance indicators requires that second-order well losses are insignificant.

Figure 4.3 shows drawdown data plotted against the well discharge rate from a step drawdown test of a fractured limestone well at Aalborg, Denmark *(Geological Survey of Denmark, 1974)*. There is a linear relationship between drawdown and discharge rate showing a linear well loss and no turbulent losses.

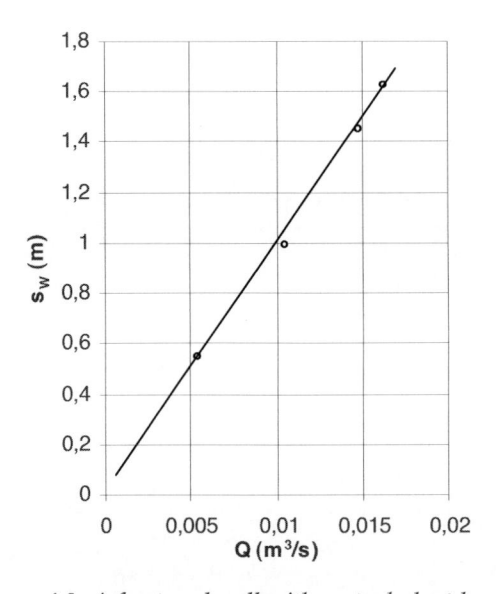

*Figure 4.3. A fractured well with no turbulent losses.*

Figure 4.4 shows the same type of plot from a fractured limestone well in Northern Zealand, Denmark *(Geological Survey of Denmark, 1975)*. Data points form a parabola, indicating turbulent losses.

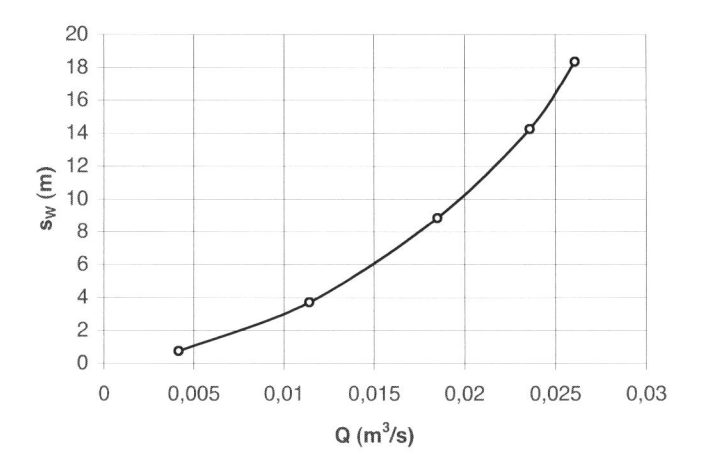

*Figure 4.4. A fractured limestone well with turbulent losses.*

Example 4.2 Well performance indicators

We use the data from Example 4.1 and get, since the drawdown at the end of the test is $s_w = 13.846$ m:

Well efficiency: $V = \dfrac{s_w - s_s}{s_w} = \dfrac{13.846 - 4.26}{13.846} = 0.69$

Clogging index: $C_i = 1/V = \dfrac{1}{0.69} = 1.44$

Clogging factor: $C_f = 1 - V = 1 - 0.69 = 0.31$

When the second-order loss is significant this should be taken into account when calculating well performance indicators. In fractured aquifers, this is a problematic task since the flow takes place in a fracture system, which is largely unknown in terms of aperture, geometry and distribution. *Ramey (1982)* has published a comprehensive study of the combined effect of skin losses and second-order losses.

*4.3 Partial penetration*
If a well partially penetrates the aquifer, the well has an additional loss because the flow velocity has a vertical component. This loss is proportional to the discharge rate and therefore behaves like a skin loss. The oil literature refers to this loss as a

# FRACTURED AQUIFERS

pseudo-skin loss determined by a pseudo-skin factor. Figure 4.5 shows a schematic of the drawdown as a function of time in a partially penetrating well.

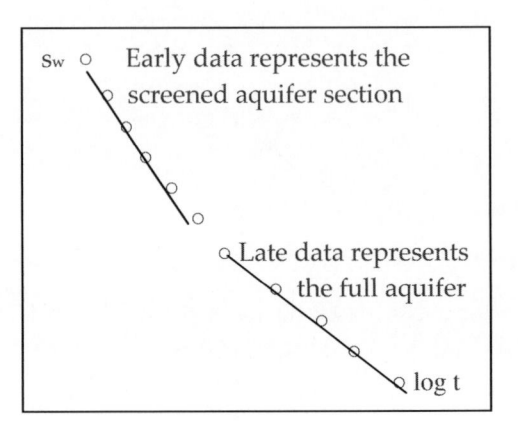

*Figure 4.5 Schematic drawdown in a partially penetrating well.*

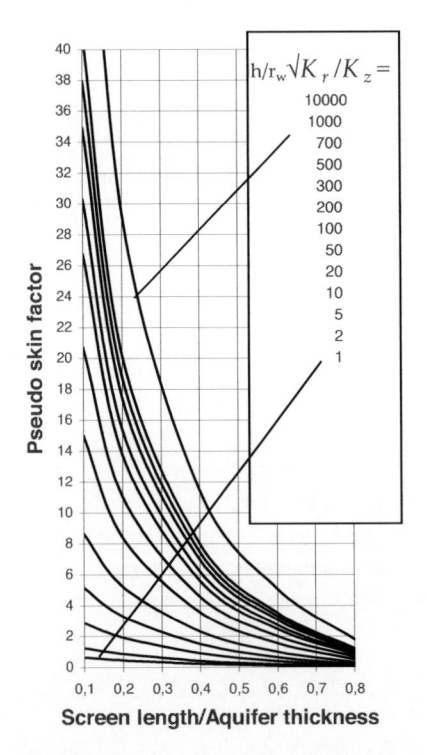

*Figure 4.6 Pseudo-skin factor for partially penetrating wells, after Brons and Marting (1961). $K_r$ and $K_z$ are horizontal and vertical conductivities respectively and h is aquifer thickness.*

Figure 4.6 shows curves used to calculate the pseudo-skin factor. If there is no clogging in the well, the figure gives the pseudo-skin factor. If there is clogging, the total skin factor is the sum of the usual skin factor $\sigma$ and the pseudo-skin factor $\sigma_p$ giving the total skin loss $s_{sT}$

$$s_{sT} = \frac{Q}{2\pi T}(\sigma + \sigma_p) \qquad (4.14)$$

## 4.4 Skin calculations from recovery data

Sometimes drawdown data is influenced by changes in discharge rate, drastic barometric pressure changes, interference from nearby discharge wells or other disturbances that may render data uncertain. In such cases, analysis of recovery data often yields better results.

Figure 4.7 shows the water level changes during the pumping and recovery periods. Mathematically, pumping continues during the recovery period which starts at time $t = t_p$ (duration of pumping), at which time $\Delta t$ (time since pumping stopped) is zero. The drawdown in the abstraction well is $s_w$, the mathematically correct recovery is $s_w''$ and the residual drawdown is $s_w'$. The correct way to plot data for analysis is to use residual drawdown data. From Equation 4.7 we get the equation for the residual drawdown

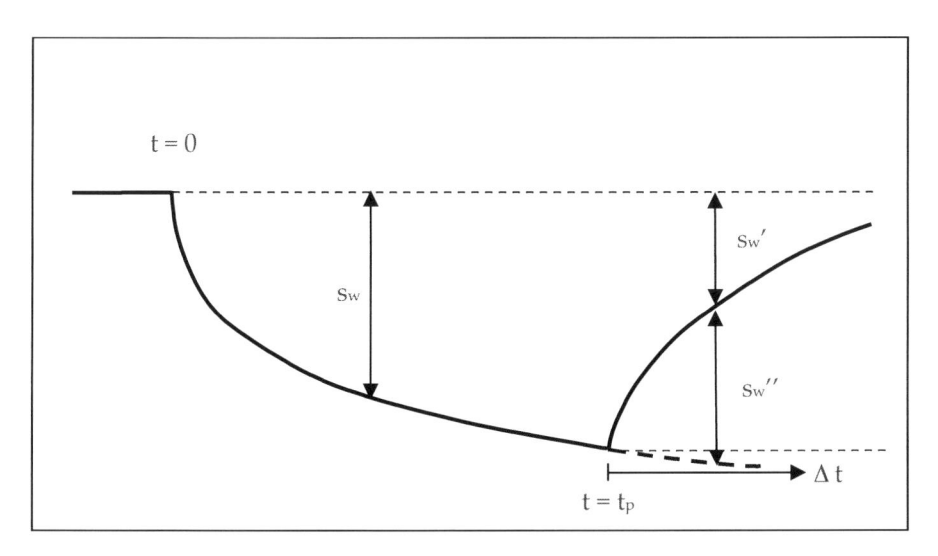

*Figure 4.7 Drawdown and recovery periods in a production well.*

$$s'_w(\Delta t) = s_w(t_p + \Delta t) - s''_w(\Delta t) =$$

$$\frac{0.183Q}{T}\left(\log\frac{2.25T(t_p + \Delta t)}{r_w^2 S} + 0.8686\sigma - \log\frac{2.25T\Delta t}{r_w^2 S} - 0.8686\sigma\right) =$$

$$\frac{0.183Q}{T}\log\frac{t_p + \Delta t}{\Delta t} \tag{4.15}$$

In this equation, the skin factor vanishes, so to determine $\sigma$ we need a different relation as an approximation. The approximate equation for the recovery after stop of pumping is

$$s''_w(\Delta t) = s_w(t_p) - s'_w(\Delta t) =$$

$$\frac{0.183Q}{T}\left(\log\frac{2.25Tt_p}{r_w^2 S} + 0.8686\sigma - \log\frac{t_p + \Delta t}{\Delta t}\right) =$$

$$\frac{0.183Q}{T}\left(\log\frac{2.25T\Delta t}{r_w^2 S} - \log\frac{t_p + \Delta t}{t_p} + 0.8686\sigma\right)$$

From this the skin factor becomes

$$\sigma = 1.1513\left(\frac{T(s_w(t_p) - s'_w(\Delta t))}{0.183Q} - \log\frac{2.25T\Delta t}{r_w^2 S} + \log\frac{t_p + \Delta t}{t_p}\right) \tag{4.16}$$

Here we can introduce a simplification by taking $\Delta t = 1$ and we get

$$s_w(t_p) - s'_w(\Delta t = 1) = \frac{0.183Q}{T}\left(\log\frac{2.25T \cdot 1}{r_w^2 S} + \log\frac{t_p}{t_p + 1} + 0.8686\sigma\right)$$

The skin factor equation now reads

$$\sigma = 1.1513\left(\frac{T(s_w(t_p) - s'_w(1))}{0.183Q} + \log\frac{t_p + 1}{t_p} - \log\frac{T}{r_w^2 S} - 2.13\right) \tag{4.17}$$

In Equation 4.17 the value of $s_w(t_p) - s'_w(1)$ is found for $t = 1$ min. If the pumping time is very long so that $t_p \gg 1$, Equation 4.17 reduces to

$$\sigma = 1.1513\left(\frac{s_w(t_p)-s'_w(1)}{\Delta s_w} - \log\frac{T}{r_w^2 S} - 2.13\right) \qquad (4.18)$$

*Example 4.3 Skin calculation from recovery data*

The example is from a recovery test on a limestone well at Hoeje Taastrup, west of Copenhagen *(Nielsen, 2004 a)*. The following data is given:

$Q = 73.8$ m³/h, $r_w = 0.1$ m, $t_p = 10,200$ min. and $S = 0.001$. Figure 4.8 and the table shows recovery data.

| (t_P + Δt)/Δt | sw-s'w (m) | (t_P + Δt)/Δt | sw-s'w (m) | (t_P + Δt)/Δt | sw-s'w (m) | (t_P + Δt)/Δt | sw-s'w (m) |
|---|---|---|---|---|---|---|---|
| 100001 | 0.04 | 3847 | 0.27 | 501 | 0.52 | 63.5 | 0.71 |
| 50001 | 0.09 | 3572 | 0.28 | 455.6 | 0.53 | 56.56 | 0.72 |
| 33334 | 0.11 | 3031 | 0.3 | 417.7 | 0.54 | 51 | 0.72 |
| 25001 | 0.13 | 2633 | 0.31 | 385.6 | 0.55 | 42.67 | 0.73 |
| 20001 | 0.15 | 2382 | 0.33 | 334.3 | 0.56 | 34.33 | 0.74 |
| 16668 | 0.15 | 2084 | 0.34 | 251 | 0.6 | 30.41 | 0.75 |
| 14287 | 0.17 | 1924 | 0.35 | 228.3 | 0.61 | 26 | 0.75 |
| 12501 | 0.18 | 1725 | 0.36 | 201 | 0.62 | 21 | 0.77 |
| 10001 | 0.19 | 1472 | 0.38 | 186.2 | 0.63 | 16.15 | 0.78 |
| 8334 | 0.21 | 1283 | 0.4 | 167.7 | 0.64 | 12.9 | 0.79 |
| 7144 | 0.22 | 1137 | 0.41 | 143.9 | 0.65 | 11.42 | 0.8 |
| 6251 | 0.23 | 1021 | 0.43 | 126 | 0.66 | 6 | 0.82 |
| 5557 | 0.24 | 834.3 | 0.45 | 112.1 | 0.67 | 4.33 | 0.84 |
| 5001 | 0.25 | 715.3 | 0.47 | 101 | 0.68 | 3.5 | 0.85 |
| 4546 | 0.26 | 626 | 0.49 | 84.33 | 0.69 | | |
| 4168 | 0.27 | 556.6 | 0.5 | 72.43 | 0.71 | | |

The transmissivity is

$$T = \frac{0.183Q}{\Delta(s_w - s'_w)} = \frac{0.183 \cdot 73.8}{0.106 \cdot 3600} = 0.0354 \text{ m}^2/\text{s}$$

Since $\frac{t_p+1}{t_p} \cong 1$, and $\Delta t = 1$ min gives $\frac{t_p+1}{1} = 10.201$, the skin factor is from Equation 4.18:

# FRACTURED AQUIFERS

$$\sigma = 1.1513\left(\frac{s_w(t_p) - s'_w(1)}{\Delta s_w} - \log\frac{T}{r_w^2 S} - 2.13\right) =$$

$$1.1513\left(\frac{0.482}{0.106} - \log\frac{0.0354}{0.1^2 \cdot 0.001} - 2.13\right) = -1.3$$

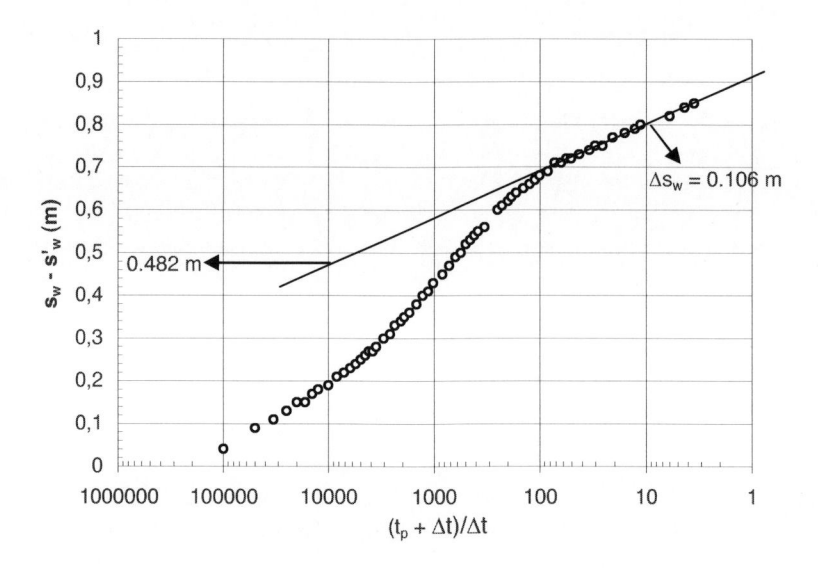

*Figure 4.6. Recovery data from a fractured limestone well.*

The skin loss becomes from Equation 4.2:

$$s_s = \frac{\sigma Q}{2\pi T} = -\frac{1.3 \cdot 73.8}{2\pi \cdot 0.0354 \cdot 3600} = -0.12 \text{ m}$$

The effective radius becomes (Equation 4.5):

$$r_{ef} = 0.1 e^{1.3} = 0.37 \text{ m}.$$

The drawdown at the end of the test was 1.09 m and the well efficiency becomes (Equation 4.11):

$$V = \frac{1.09 + 0.12}{1.09} = 111 \%$$

This well is stimulated due to the fractures.

## 4.5 Well bore storage

When the discharge rate is small, early drawdown and recovery data may show the effect of well bore storage. During this period the drawdown or recovery in the well is smaller than predicted by the Theis equation, see Figure 2.1. The time required to empty the water in the casing for the aquifer to take control is

$$t = \frac{\pi \left( r_c^2 - r_{rm}^2 \right) s_w}{Q} \tag{4.19}$$

Here, $r_c$ is the casing radius, $r_{rm}$ is the rising main radius, and $s_w$ is the drawdown for which the well bore storage period ends. The well bore storage period in a fractured well looks like the early fracture response. To verify the effect of well bore storage, we use Equation 4.19 to calculate the time at which the well bore storage effect vanishes. This time should match the time observed from the data plot.

In a fractured well, storage in fractures near the well may extend the storage period considerably compared to the period calculated from Equation 4.19.

# 5. Pressure-dependent transmissivity

In a confined aquifer, partly the water and partly the rock matrix carries the weight of the overburden. During pumping, the water pressure reduces and weight transfers to the rock matrix, which consolidates. In a sand aquifer, consolidation is insignificant and hardly noticeable in terms of transmissivity changes. This is not the case in fractured aquifers, because the flow rate in fractures depends strongly on the fracture aperture, and even small pressure changes or drawdowns may change transmissivities significantly.

Pumping from an aquifer with pressure dependent transmissivity results in drawdowns increasing with time beyond those given by the Theis solution. Since the transmissivity depends on the drawdown the governing differential equation becomes non-linear and solving it is not a trivial task. There does not seem to be much literature on the subject. The work included here is that of *Sorensen (1981)*.

## 5.1 Sorensen's model
### 5.1.1 Drawdown data

Sorensen's model assumes the following mathematical relationship between drawdown $s$ and transmissivity

$$T = T_0 e^{-\alpha s}, \text{ where} \tag{5.1}$$

$T_0$ is the initial transmissivity ($s = 0$), and $\alpha$ is a constant. Differentiation of Equation 5.1 yields

$$\frac{dT}{ds} = -\alpha T_0 e^{-\alpha s} = -\alpha T,$$

i.e. $\alpha$ is the ratio between the pressure derivative of transmissivity and the transmissivity. With the following dimensionless parameters

Dimensionless drawdown $s_D = \dfrac{4\pi T_0 s}{Q}$

Dimensionless time $u_0 = \dfrac{r^2 S}{4 T_0 t}$ and $\beta = \dfrac{\alpha Q}{4\pi T_0}$,

Sorensen found the following approximate solution for the drawdown, $(u_0 < 0.01)$

# FRACTURED AQUIFERS

$$s_D = -\frac{1}{\beta}\ln\left(1-\beta\ln\frac{0{,}561}{u_0(1+\beta)}\right) \tag{5.2}$$

Written in field notation Equation 5.2 reads

$$s = -\frac{1}{\alpha}\ln\left(1-\frac{\alpha Q}{4\pi T_0}\ln\frac{2{,}25T_0 t}{r^2 S\left(1+\frac{\alpha Q}{4\pi T_0}\right)}\right) \tag{5.3}$$

It appears from Equation 5.3 that the parameter $\alpha$ controls the drawdown. Figure 5.1 shows a plot of Equation 5.2 for the abstraction well ($r = r_w$).

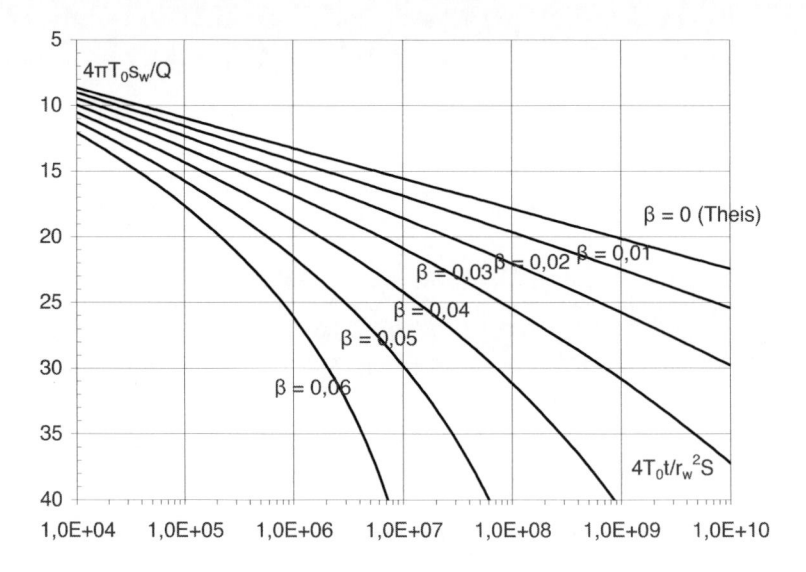

*Figure 5.1 Dimensionless drawdown versus dimensionless time in an aquifer with pressure-dependent transmissivity.*

For $\beta = 0$ we get $\alpha = 0$, and there is no pressure dependency. The drawdown follows Theis' equation

$$s = \frac{Q}{4\pi T_0}W\left(\frac{r_w^2 S}{4T_0 t}\right)$$

## 5.1.2 Recovery data

When the governing differential equation is linear, the superposition principle is valid and we get the same hydraulic parameters whether drawdown or recovery data is used. This is not the case when the differential equation is non-linear. An exact solution of the recovery case is not available, but Sorensen presented an approximate solution assuming a certain shape of the cone of depression during recovery. Figure 5.2 shows this solution.

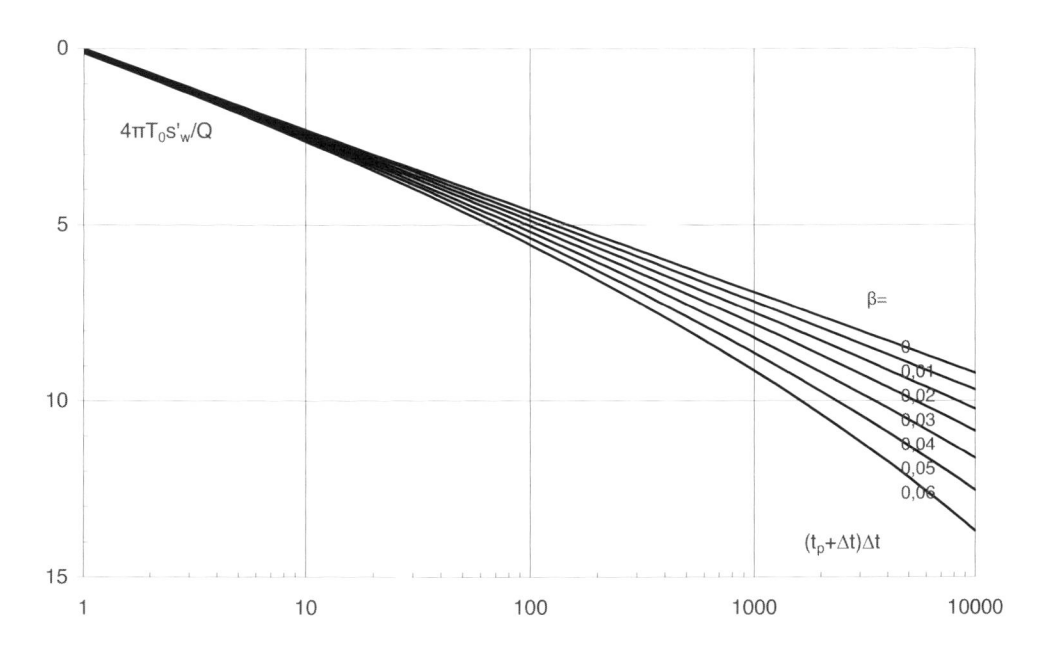

*Figure 5.2 Dimensionless recovery curves.*

Determination of hydraulic parameters is not trivial and requires an iterative process. It has not been possible to identify drawdown or recovery data matching Sorensen's model and therefore, we do not present the procedure here. However, Sorensen's model implies important conclusions.

## 5.1.3 Discussion

Sorensen's model does not make any assumptions about what kind of fractures exist in the aquifer and there is no mention of dual porosity. The model assumes homogeneous aquifer conditions and therefore gives no information about the fracture system, which is the case for other models presented in subsequent chapters.

The reason that no cases found match the type-curves for an aquifer with pressure-dependent transmissivity may be that the condition in Equation 5.1 is too strict. It prescribes a transmissivity that tends towards zero as the drawdown increases. A more realistic condition is to allow the transmissivity to approach matrix transmissivity $T_m$ according to a condition of the type

$$T = T_m + (T_0 - T_m)e^{-\alpha s}$$

In this way, the drawdown curve would approach a straight line in a semi-log plot, from which matrix transmissivity and bulk storage could be determined using the Theis equation.

An important feature, which appears when comparing Figures 5.1 and 5.2, is that the log cycle slope during drawdown is considerably larger than during recovery. This is a good indication that the aquifer has a pressure-dependent tranmissivity. The physical explanation for this is that at the end of the drawdown period the transmissivity is low, but after stop of pumping the transmissivity increases rapidly resulting in a lower log cycle slope.

When calculating future drawdowns in this type of aquifer we shall use transmissivity values determined during the drawdown period. Using values determined from recovery data will result in under estimation of drawdowns.

The phenomenon of pressure dependent transmissivity exists only in confined aquifers, not in phreatic aquifers.

## 6. The dual-porosity aquifer

The dual porosity rock formation consists of units of porous rock material separated by fractures. It is obvious that no matter how many wells are available, a detailed description of fracture geometry is not possible. To evaluate formation constants, it is necessary to prepare models that include a simplified or ideal fracture system. The usefulness of such models will depend on how accurately the simplified fracture system resembles the actual one. As we shall see, the mathematical models available are able to give an accurate description of ground water flow towards wells in fractured rock formations in spite of the simplifications made. The oil literature refers to the fractured aquifer as a *naturally fractured reservoir* as opposed an artificially fractured reservoir.

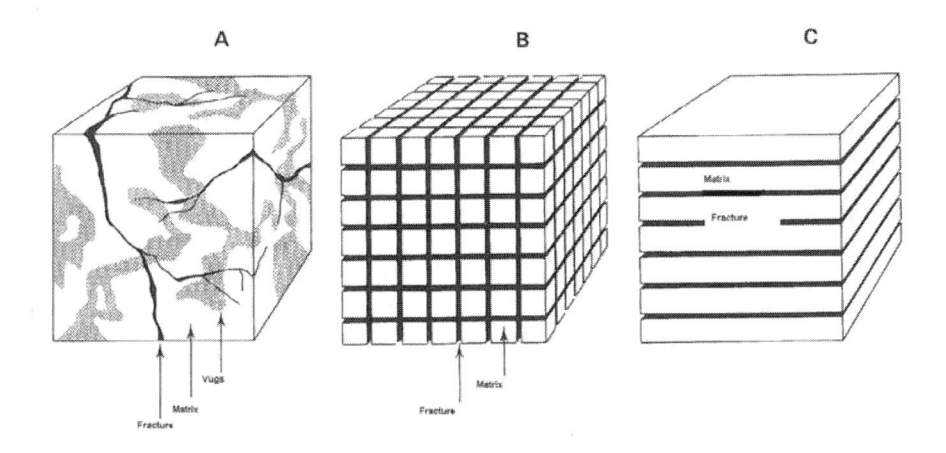

*Figure 6.1 Fractured rock formation, from Kruseman and de Ridder (2001).*
    *A. Naturally fractured rock formation.*
    *B. Idealized fractured rock formation with cubic matrix blocks.*
    *C. Idealized horizontally fractured rock formation.*

A number of mathematical models of fractured aquifers exist as described in this chapter. The same fundamental assumptions on how flow in the formation takes place apply to them all:

- Ground water enters the well through fractures only.

- The fractures form a secondary porosity system with high hydraulic conductivity and low storage capability.

- The matrix has low hydraulic conductivity and high storage capability. There is no flow of ground water from matrix to the well, but matrix releases water to fractures during pumping due to the lower pressure in fractures. The flow rate is proportional to the pressure difference between fractures and matrix. This inter-porosity flow is stationary in some models and non-stationary in others. The ratio of fracture to matrix hydraulic conductivities is typically in the range $10^3 - 10^5$, and the ratio of fracture to matrix storage is typically $10^{-3} - 10^{-1}$.

The assumptions behind the Theis solution are used, i.e. the aquifer is homogeneously fractured, the well fully penetrates the aquifer, the aquifer has infinite areal extent, etc.

## 6.1 Warren and Root

*Warren and Root (1963)* published the first practical model for evaluation of formation constants in a naturally fractured aquifer. Their work is a classic and the base for most subsequent work on naturally fractured aquifers.

Warren and Root solved the equation for drawdown in the pumped well alone and introduced three parameters characterizing the aquifer properties and the ground water flow during pumping:

$$\omega = \frac{S_f}{S_f + S_m} \tag{6.1}$$

$$\lambda = \alpha r_w^2 \frac{K_m}{K_f} \tag{6.2}$$

$$\alpha = \frac{4n(n+2)}{l^2} = \frac{4n(n+2)}{h_m^2} \quad \text{, where} \tag{6.3}$$

$\omega$ is the ratio of fracture storage to bulk storage, $S_f$ is fracture storage, $S_m$ is matrix storage, $\lambda$ is the inter-porosity flow factor controlling the flow from matrix to fractures, $K_m$ is matrix hydraulic conductivity, $K_f$ is fracture hydraulic conductivity, $l$ is the characteristic matrix block or layer length, for which we here use the length of the matrix block or thickness of the matrix layer $h_m$, and n is the number of normal sets of fractures, i.e. $n = 1$ for the layer model and $n = 3$ for the blocks model.

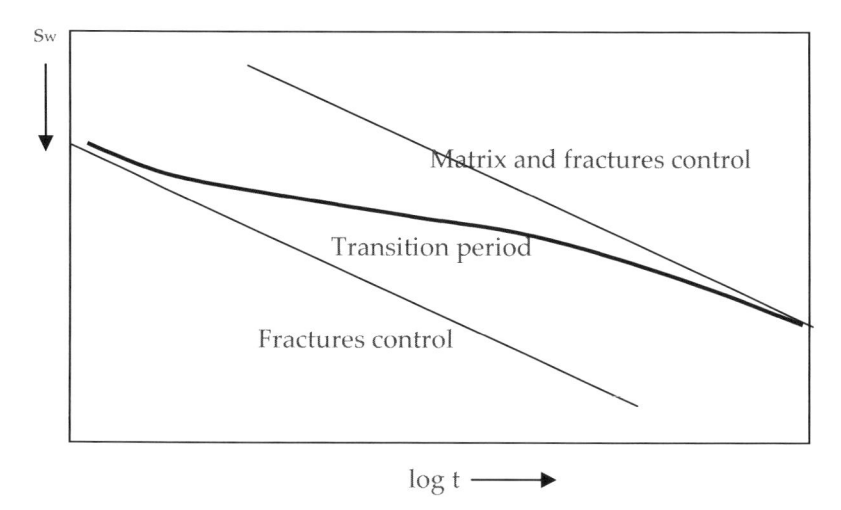

*Figure 6.2 Schematic drawdown in a dual-porosity aquifer according to Warren and Root's model.*

The essential property of Warren and Root's model is the realization that the drawdown in a dual-porosity aquifer develops an early and a late phase. Early data reveals only fracture properties and produces a straight line in a semi-log plot. Late data reveals fracture as well as matrix properties and represents the homogeneous aquifer. Again, this data falls on a straight line in a semi-log plot. These two straight lines are Theis solutions and parallel since they both define the fracture transmissivity $T_f$. A transition phase connects the early and late lines. The flow from matrix to fractures controls the transition period. During the transition phase, the water released from matrix may equal the discharge and a pseudo-steady state develops for some time (minutes, hours). Figure 6.2 shows a schematic of the drawdown as a function of time. Warren and Root's model assumes stationary flow between matrix and fractures, i.e. time is not involved in the inter-porosity flow condition.

Warren and Root's approximate solution for the drawdown reads

$$s_w = \frac{Q}{4\pi T_f}\left(2.3\log\frac{2.25T_f t}{r_w^2(S_f+S_m)} - W\left(\frac{\lambda T_f t}{r_w^2 S_f(1-\omega)}\right) + W\left(\frac{\lambda T_f t}{r_w^2(S_f+S_m)(1-\omega)}\right)\right) \quad (6.4)$$

where the condition

$$t_D = \frac{T_f t}{r_w^2 (S_f + S_m)} > 100 \tag{6.5}$$

applies. Water supply wells normally fulfill this condition and Equation 6.4 is therefore usually valid for all values of $\lambda$ and $\omega$.

For small values of time, Equation 6.4 reduces to

$$s_w = \frac{0.183 Q}{T_f} \log \frac{135 T_f t}{r_w^2 S_f} \tag{6.7}$$

For large values of time we get

$$s_w = \frac{0.183 Q}{T_f} \log \frac{135 T_f t}{r_w^2 (S_f + S_m)} \tag{6.8}$$

We recognize Equations 6.7 and 6.8 as the Jacob – Cooper equation (Equation 2.3), only the physical parameters are different.

For the residual drawdown Warren and Root found the following equation,

$$s_w' = \frac{Q}{4\pi T_f} \left( 2.3 \log \frac{t_p + \Delta t}{\Delta t} + W\left(\frac{\lambda T_f \Delta t}{r_w^2 S_f (1-\omega)}\right) - W\left(\frac{\lambda T_f \Delta t}{r_w^2 (S_f + S_m)(1-\omega)}\right) \right) \tag{6.9}$$

for $\dfrac{T_f \Delta t}{r_w^2 (S_f + S_m)} > 100\omega$ if $\lambda \ll 1$, or $\dfrac{T_f \Delta t}{r_w^2 (S_f + S_m)} > 100 - \dfrac{1}{\lambda}$ if $\omega \ll 1$.

The vertical displacement $\delta s_w$ between the early and late line may be found by subtracting Equation 6.8 from Equation 6.7, see Figure 6.3:

$$\delta s_w = \frac{0.183 Q}{T_f} \left( \log \frac{135 T_f t}{r_w^2 S_f} \cdot \frac{r_w^2 (S_f + S_m)}{135 T_f t} \right) = -\Delta s_w \log \omega$$

From this we get

$$\omega = 10^{-\delta s_w / \Delta s_w} \tag{6.10}$$

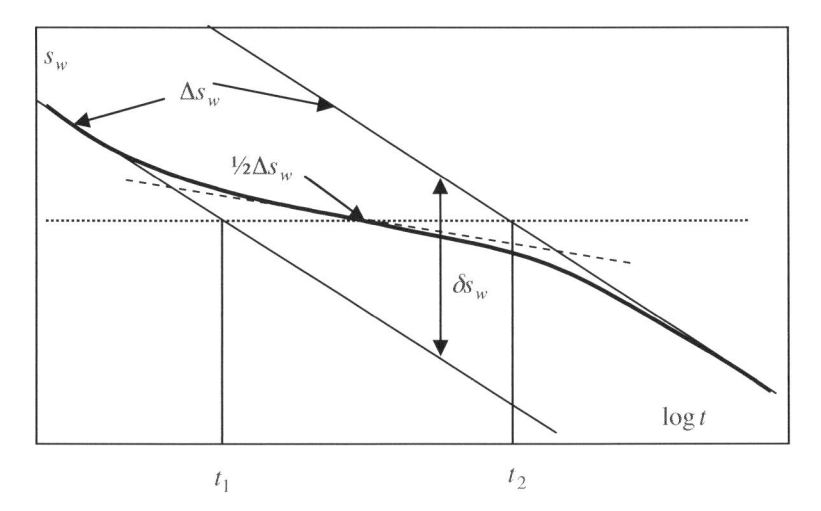

*Figure 6.3 Evaluation of aquifer parameters.*

The horizontal displacement between the lines also defines $\omega$ :

$$\omega = \frac{t_1}{t_2} \tag{6.11}$$

*Bourdet and Gringarten (1980)* demonstrated a method for calculating $\lambda$ by drawing the horizontal line through the inflection point (the mid point of the transition curve) and note the times $t_1$ and $t_2$, where the horizontal line intersects the early and late straight lines. They found the following equations for drawdown data

$$\lambda = \frac{r_w^2 S_f}{\gamma T_f t_1} = \frac{r_w^2 (S_f + S_m)}{\gamma T_f t_2} \tag{6.12}$$

where $\gamma = 1.78$ (the exponential of Euler's constant). The equation for recovery data reads

$$\lambda = \frac{r_w^2 S_f (t_p + \Delta t_1)}{\gamma T_f t_p \Delta t_1} = \frac{r_w^2 (S_f + S_m)(t_p + \Delta t_2)}{\gamma T_f t_p \Delta t_2} \tag{6.13}$$

With these equations and including the skin calculation, we are now in principle able to determine the well and aquifer parameters that characterize a dual-porosity rock formation, except for storage values, the evaluation of which requires monitoring well data. It often happens that only the late part of the transition curve and the late data representing the homogeneous aquifer are seen, not the early data. In this case, only the fracture transmissivity can be determined. Whether or not we see all three phases of the drawdown curve depends on several factors.

In confined aquifers with small storage coefficients, the cone of depression develops rapidly and the early data may not appear because the transition happens within a fraction of a minute. Often, the effect of well bore storage masks the early data. In phreatic aquifers, there is a greater chance to obtain a larger part of the drawdown curve.

Another factor is the size of matrix blocks or layers in combination with the matrix hydraulic conductivity. For small blocks, the pressure reduction in matrix happens fast and we may see only part of the transition curve, especially for large matrix hydraulic conductivities. For thick blocks and small matrix hydraulic conductivities the pressure difference between fractures and matrix may last for some time, and we may see a large part or the whole drawdown curve, even obtain a pseudo-steady state.

It is worth noting that the dual-porosity effect does not reveal itself on the piezometric surface under natural conditions. This is also the case for vertical and horizontal single fractures, which will be introduced later. The reason for this is that under undisturbed conditions there is pressure equilibrium between fractures and matrix.

To see the influence of fractures it is necessary to stress the aquifer by pumping. This has a bearing on contaminant surveys. Ground water flows in fractures and so will contaminants. Therefore, particle paths may be very different from the direction given by the apparent gradient of the piezometric surface. Contaminant surveys in fractured aquifers have little chance of yielding satisfactory results without testing the aquifer hydraulically. Monitoring wells will provide valuable data on fracture extension and direction.

*Uldrich and Ershagi (1979)* developed a graphical method to determine $\omega$ and $\lambda$ even if the transition period is not seen. Their method is speculative and uncertain and will not be included here.

*Serra, Reynolds and Raghaven (1983)* with a starting point in Warren and Root's solution further analyzed the equations. They found that the slope of the transition curve at the inflection point is ½$\Delta s_w$, see Figure 6.3. The line where the slope is

½Δ$s_w$ also appears in homogeneous aquifers influenced by a barrier boundary. Actually, it is common to see such drawdown curves from fractured aquifers interpreted as being influenced by a barrier boundary.

The transition curve between early and late data may reach a pseudo-steady state, which may last for hours. The drawdown curve resembles a drawdown curve from a leaky aquifer. If leakage is the interpretation preferred, this will grossly under estimate the forecast of future drawdowns. It is a rule of thumb that a well test in a fractured aquifer should continue at least until data from the pumped well shows the behavior of the homogeneous aquifer.

6.1.1 The significance of $\lambda$ and $\omega$

Figure 6.4 shows the drawdown in a well according to Warren and Root's model. For large values of $\lambda$ the transition phase ends quickly and often does not show. This is the case if fracture transmissivity is large or matrix transmissivity is small. The limiting case $\lambda = \infty$ reflects that the fracture transmissivity is zero, there is no dual-porosity and we only see matrix properties.

If $\omega$ is small, i.e. fracture storage is small compared to bulk storage, the horizontal displacement between the early and late straight lines is large. The limiting case $\omega = 1$ corresponds to fracture porosity only, i.e. the matrix is non-permeable. Very small values of $\omega$ may result in pseudo-steady state and a late transition to homogeneous aquifer behavior, while large values of $\omega$ result in a short transition period.

Figure 6.5 shows the recovery in a dual-porosity aquifer. We may not always see the two parallel lines. The line representing the limiting case $\omega = 0$ (no or very small fracture porosity) moves upward in the diagram and approaches the initial pressure as $\lambda$ decreases (the figure to the right). This shows that recovery happens almost instantaneously if $\lambda$ is very small.

*6.2 de Swaan*

*de Swaan (1976)* presented an analytical method for evaluation of hydraulic parameters in a dual-porosity aquifer. He was not concerned with the transition curve but treated only early and late data, and therefore did not involve the parameters $\lambda$ and $\omega$ in the analyses.

For small times the equation for drawdown in the pumped well is

$$s_w = \frac{2.3Q}{4\pi T}\log\frac{2.25\eta_f t}{r_w^2} \qquad (6.14)$$

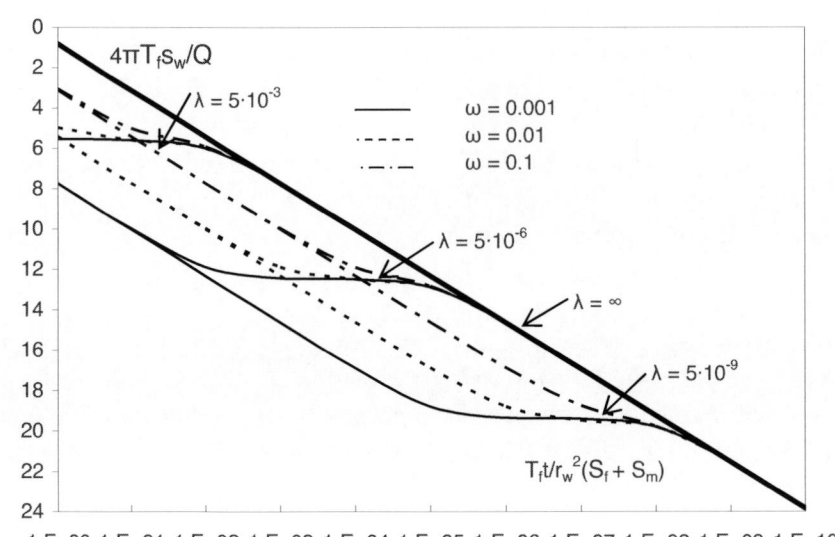

*Figure 6.4 Dimensionless drawdown as a function of dimensionless time for different values of $\lambda$ and $\omega$.*

where $\eta_f = \dfrac{T_f}{S_f}$ is the fracture diffusivity. This equation is identical to Warren and Root's Equation 6.7. For large times de Swaan used a similar equation but with a modified diffusivity. For the layer model, see Figure 6.6,

$$\eta_L = 1\bigg/\left(\frac{1}{\eta_f}+\frac{K_m h_m}{K_f h_f \eta_m}\right) = 1\bigg/\left(\frac{1}{\eta_f}+\frac{T_m}{T_f \eta_m}\right) \qquad (6.15)$$

where $h_f$ is the fracture aperture and $\eta_m = \dfrac{T_m}{S_m}$ is the matrix diffusivity.

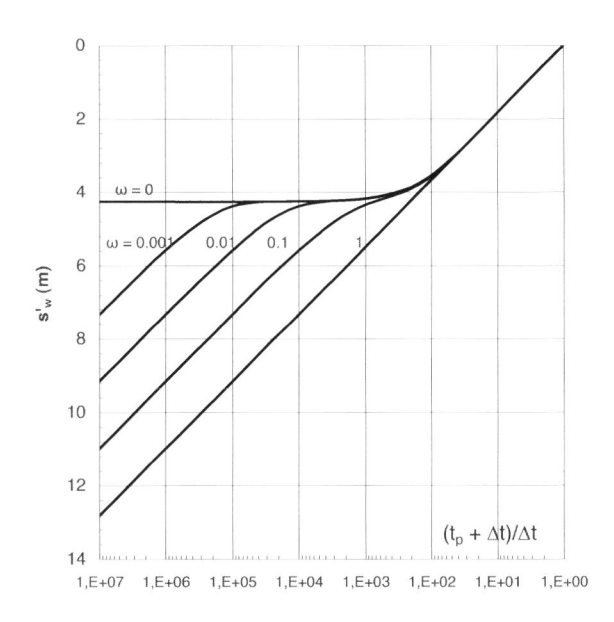

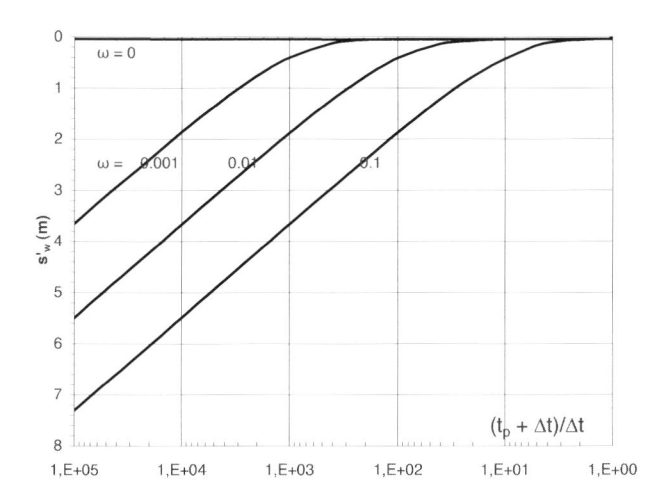

*Figure 6.5 Theoretical recovery curves from a fractured well.*
$T_f = 0.001\, m^2/s$, $\lambda = 10^{-4}$ *(top)*, $S_f + S_m = 0.1$,
$r_w = 0.1\, m$, $t_p = 20{,}000\, min$, $Q = 0.01\, m^3/s$.
$\lambda = 5 \cdot 10^{-7}$ *(bottom), other parameters unchanged.*

# FRACTURED AQUIFERS

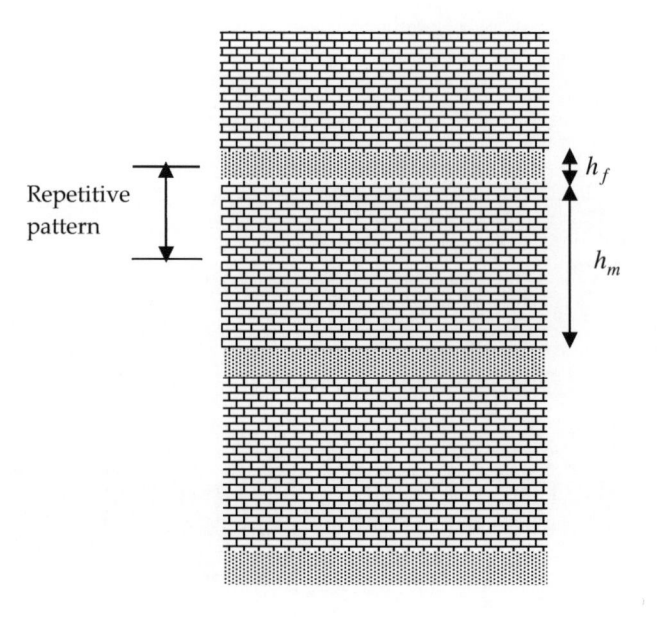

*Figure 6.6 Horizontally fractured aquifer. After de Swaan (1976).*

For blocks we have

$$\eta_B = 1 / \left( \frac{1}{\eta_f} + \frac{2 K_m r_m}{3 K_f h_f \eta_m} \right) \qquad (6.16)$$

Equation 6.16 is valid for spherical blocks with radius $r_m$ assuming that they behave like cubic blocks.

Finally, de Swaan derived an equation for the diffusivity for blocks of arbitrary shape and matrix storage:

$$\eta_K = 1 / \left( \frac{1}{\eta_f} + \frac{\overline{S_m}}{K_f h_f} \right) \qquad (6.17)$$

where $\overline{S_m}$ is an average storage, which considers block geometry.

If we write out the diffusivity in Equation 6.15, we get Warren and Root's Equation 6.8. The use of Equations 6.16 and 6.17 requires knowledge of the micro-geometry of the rock formation. Only laboratory analysis of core samples can provide that information, and this is standard procedure in the oil industry. This is not customary in the water supply industry, and de Swaan's method is less useful for water wells, but we conclude that his work is a confirmation of that of Warren and Root.

## 6.3 Kazemi

*Kazemi (1969)* developed a numerical model of a dual-porosity aquifer with non-steady inter-porosity flow. Kazemi's model matches exactly Warren and Root's for early and late data, but the transition period differs. Since the method is fully numerical it is difficult to use in practice and Kazemi's model does not contribute essentially new knowledge compared to Warren and Root's method.

*Kazemi, Seth and Thomas, (1969)* published an analytical solution for the drawdown in a monitoring well in a dual-porosity aquifer. With the usual notation their solution reads

$$s(r,t) = \frac{Q}{4\pi T_f}\left(2.3\log 2.25 t_D - W\left(\frac{\lambda t_D}{\omega(1+\omega)}\right) + W\left(\frac{\lambda t_D}{(1-\omega)}\right)\right) \qquad (6.18)$$

where dimensionless time is $t_D = \dfrac{T_f t}{r^2(S_f + S_m)}$. Equation 6.18 is exactly Warren and Root's Equation 6.4 with the well radius $r_w$ replaced by the distance $r$ to the monitoring well. The use of Equation 6.18 requires that $t_D = \dfrac{T_f t}{r^2(S_f + S_m)} > 100$. For practical purposes, the condition may be weakened to $t_D > 20$. Equation 6.18 plotted on log-log paper produces type curves with $\lambda$ and $\omega$ as parameters. However, the equation is an approximation and the method fails for small times or large distances. Other published methods do not have this limitation.

Example 6.1 Fractured well

Data is from a fractured limestone well at Hoeje Taastrup west of Copenhagen, *(Nielsen, 2006 b)*. The following data is given:

$Q = 15.1$ m³/h, $r_w = 0.1$ m, and from monitoring wells $S_f + S_m = 0.021$. The first data points, up to about 0.5 min, show the effect of well bore and fracture storage. We can identify the early and late straight lines as shown in Figure 6.7. We get

$$T_f = \frac{0.183Q}{\Delta s_w} = \frac{0.183 \cdot 15.1}{0.066 \cdot 3600} = 0.0116 \text{ m}^2/\text{s}$$

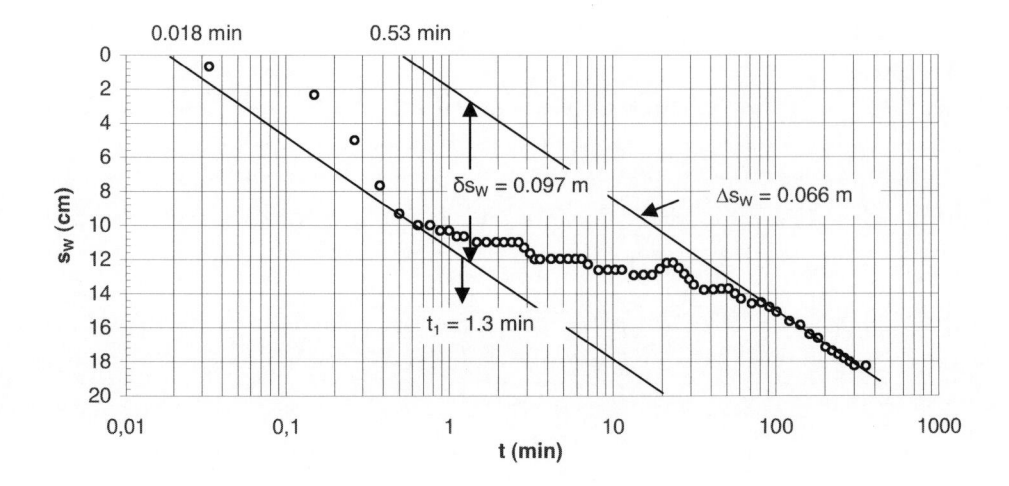

*Figure 6.7 Drawdown data from a fractured well.*

| t (min) | s_w (cm) | t (min) | s_w (cm) | t (min) | s_w (cm) | t (min) | s_w (cm) | t (min) | s_w (cm) |
|---|---|---|---|---|---|---|---|---|---|
| 0.03 | 0.67 | 1.95 | 10.99 | 7.03 | 12.29 | 29.37 | 13.16 | 141.37 | 15.84 |
| 0.15 | 2.33 | 2.17 | 10.99 | 8.17 | 12.62 | 21.37 | 13.48 | 161.37 | 16.39 |
| 0.27 | 5 | 2.42 | 10.98 | 9.32 | 12.61 | 36.37 | 13.79 | 181.37 | 16.61 |
| 0.38 | 7.66 | 2.67 | 10.98 | 10.37 | 12.61 | 41.37 | 13.76 | 201.37 | 17.16 |
| 0.5 | 9.33 | 2.88 | 11.32 | 11.37 | 12.6 | 46.37 | 13.73 | 221.37 | 17.37 |
| 0.65 | 10 | 3.13 | 11.65 | 13.37 | 12.92 | 51.37 | 13.7 | 241.37 | 17.59 |
| 0.77 | 9.99 | 3.35 | 11.98 | 15.37 | 12.91 | 56.37 | 14 | 261.37 | 17.8 |
| 0.88 | 10.33 | 3.6 | 11.98 | 17.37 | 12.9 | 61.37 | 14.31 | 281-37 | 18.02 |
| 1 | 10.33 | 4.18 | 11.97 | 19.37 | 12.55 | 71.37 | 14.58 | 301.37 | 18.24 |
| 1.12 | 10.66 | 4.77 | 11.97 | 21.37 | 12.21 | 81.37 | 14.52 | 354.37 | 18.26 |
| 1.23 | 10.66 | 5.33 | 11.97 | 23.37 | 12.2 | 91.37 | 14.8 | | |
| 1.48 | 10.99 | 5.9 | 11.96 | 25.37 | 12.52 | 101.37 | 15.07 | | |
| 1.7 | 10.99 | 6.45 | 11.96 | 27.37 | 12.84 | 121.37 | 15.62 | | |

From Equation 6.11, $\omega = 0.018/0.53 = 0.034$ or from Equation 6.12

$$\omega = 10^{-\delta s_w / \Delta s_w} = 10^{-0.097/0.066} = 0.034$$

For the fracture storage we get

$$S_f = \omega(S_f + S_m) = 0.034 \cdot 0.021 = 0.000712$$

The inter-porosity flow factor is from Equation 6.12:

$$\lambda = \frac{r_w^2 S_f}{\gamma T_f t_1} = \frac{0.1^2 \cdot 0.000712}{1.78 \cdot 0.0116 \cdot 1.3 \cdot 60} = 0.00000442$$

The value of $\omega$ shows that 3.4 % of the ground water flows in fractures. The value of $\lambda$ is relatively high and indicates a quite porous matrix.

The skin factor becomes

$$\sigma = 1.1513 \left( \frac{s_w}{\Delta s_w} - \log t - \log \frac{T_f}{r_w^2 (S_f + S_m)} - 2.13 \right) =$$

$$1.1513 \left( \frac{0.15}{0.066} - \log 100 - \log \frac{0.0116}{0.01 \cdot 0.021} - 2.13 \right) = -4.14$$

Here we have used $t = 100$ min. and $s_w = 0.15$ m. The effective radius becomes

$$r_{ef} = r_w e^{-\sigma} = 0.1 \cdot e^{4.14} = 6.28 \text{ m and the skin loss}$$

$$s_s = \frac{\sigma Q}{2\pi T_f} = -\frac{4.14 \cdot 15.1}{2\pi \cdot 0.0116 \cdot 3600} = -0.238 \text{ m}$$

At the end of the test the drawdown $s_w = 0{,}184$ m and the well efficiency is

$$V = \frac{s_w - s_s}{s_w} = \frac{0.184 + 0.238}{0.184} = 229 \text{ \%}$$

This well is highly stimulated. The fractures reduce the drawdown and increase the well radius. Well efficiencies over 100 % is a certain indication of a fractured well.

With the values determined above, we can write the equation for the drawdown in meters from Warren and Root's Equation 6.4 and include the skin loss. Note that an adjusted value of $\lambda = 0.00000492$ is used to obtain a better data fit.

$$s_w = \frac{Q}{4\pi T_f}\left(2.3\log\frac{2.25T_f t}{r_w^2(S_f + S_m)} - W\left(\frac{\lambda T_f t}{r_w^2 S_f(1-\omega)}\right) + W\left(\frac{\lambda T_f t}{r_w^2(S_f + S_m)(1-\omega)}\right)\right) =$$

$$\frac{15.1}{4\pi \cdot 0.0116 \cdot 3600}(2.3\log\frac{135 \cdot 0.0116 \cdot t}{0.1^2} - W\left(\frac{0.00000492 \cdot 0.0116 \cdot t \cdot 60}{0.1^2 \cdot 0.000712 \cdot 0.966}\right)$$

$$+ W\left(\frac{0.00000492 \cdot 0.0116 \cdot t \cdot 60}{0.1^2 \cdot 0.021 \cdot 0.966}\right)) - 0.238 =$$

$$0.0288 \cdot (2.3\log(7457 \cdot t) - W(0.498 \cdot t) + W(0.0169 \cdot t)) - 0.238$$

Figure 6.8 shows this equation plotted against data.

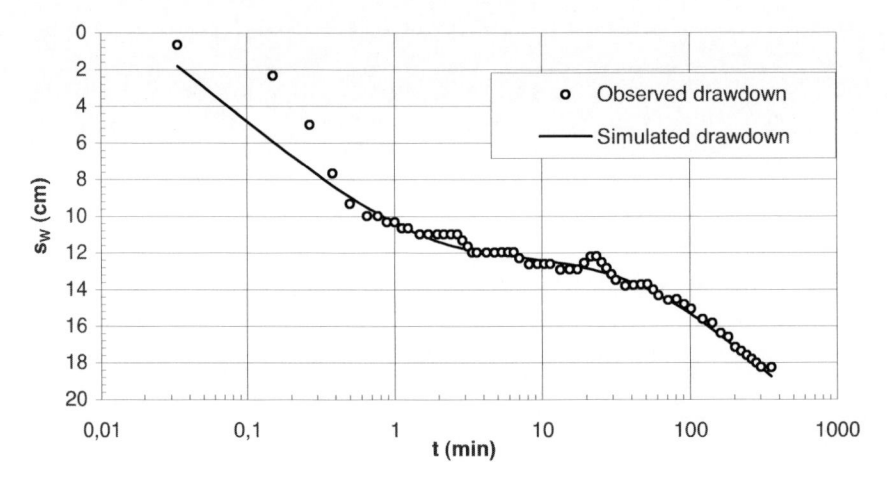

*Figure 6.8 Observed and simulated drawdown for Example 6.1.*

### 6.4 Bourdet and Gringarten

*Bourdet and Gringarten (1980)* presented a general solution for the drawdown in a monitoring well in a dual-porosity aquifer, to which Kazemi's Equation 6.18 is an approximation. Their solution written symbolically reads

$$s = \frac{Q}{4\pi T_f} F(u^*, \lambda, \omega) \qquad (6.19)$$

where

$$u^* = \frac{T_f t}{(S_f + \xi S_m)} \qquad (6.20)$$

$\xi$ is zero for early data (fractures dominate) and one for late data (homogeneous aquifer behavior) in a layered aquifer. For small times, Equation 6.19 reduces to

$$s = \frac{Q}{4\pi T_f} W\left(\frac{r^2 S_f}{4 T_f t}\right) \quad (6.21)$$

For large times we get

$$s = \frac{Q}{4\pi T_f} W\left(\frac{r^2 (S_f + S_m)}{4 T_f t}\right) \quad (6.22)$$

Equations 6.21 and 6.22 are similar to the early and late solutions the pumped well. Figure 6.9 shows a schematic of the drawdown curves. It appears from the figure that the parameter $\lambda$ determines the drawdown at which pseudo-steady state develops independently of early and late drawdowns. This was also the case for Warren and Root's solution. The parameter $\omega$, as before, determines the horizontal displacement between early and late data Theis curves. If $\omega$ is large, pseudo-steady state does not happen. Instead, the drawdown curve has an inflection point.

Bourdet and Gringarten showed for small values of $\lambda$ that the value where the drawdown curve is horizontal or has an inflection point is determined by the equation,

$$s = \frac{Q}{2\pi T_f} K_0\left(\sqrt{\lambda}\right) \Rightarrow s = \frac{2.3 Q}{4\pi T_f} \log \frac{1.26}{\lambda} \Rightarrow$$
$$\lambda = 1.26 / 10^{4\pi s T_f / 2.3 Q} \quad (6.23)$$

where $K_0(x)$ is the modified Bessel function of the second kind and zero order. Figure 6.10 shows the graph of this function.

For small $\omega$ values the curves in Figure 6.9 resemble those seen in leaky artesian aquifers. If these curves develop in fractured aquifers it is important to let the test continue until the late Theis curve (the homogeneous aquifer) develops.

It is possible to prepare type curves for monitoring wells by means of Bourdet and Gringarten's model, but that would require a large number of type-curve sheets to cover the practical range of $\lambda$ and $\omega$ values. *Kruseman and de Ridder (2001)* suggested a practical approach.

FRACTURED AQUIFERS

The method applies the type-curve procedure. First fracture parameters $T_f$ and $S_f$ are determined by matching the Theis curve to early data and reading from the data sheet the Match Points $s_{MP}$ and $t_{MP}$ using type-curve Match Points $W(u*) = 1/u* = 1$. We calculate fracture transmissivity and storage from the equations

$$T_f = \frac{Q}{4\pi s_{MP}} \tag{6.24}$$

$$S_f = \frac{240 T_f t_{MP}}{r^2}, \tag{6.25}$$

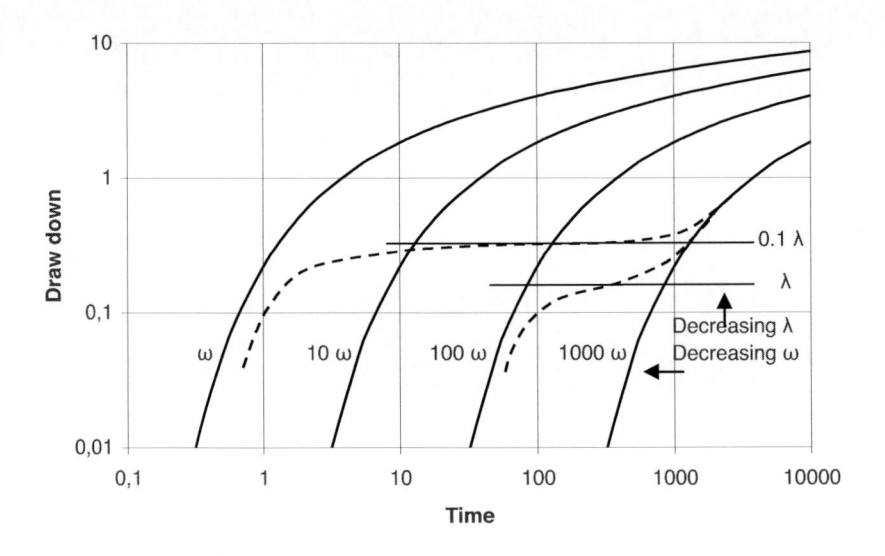

*Figure 6.9 Schematic of the drawdown in a monitoring well in a dual-porosity aquifer. After Kruseman and de Ridder (2001). Broken lines are data, other curves are Theis curves.*

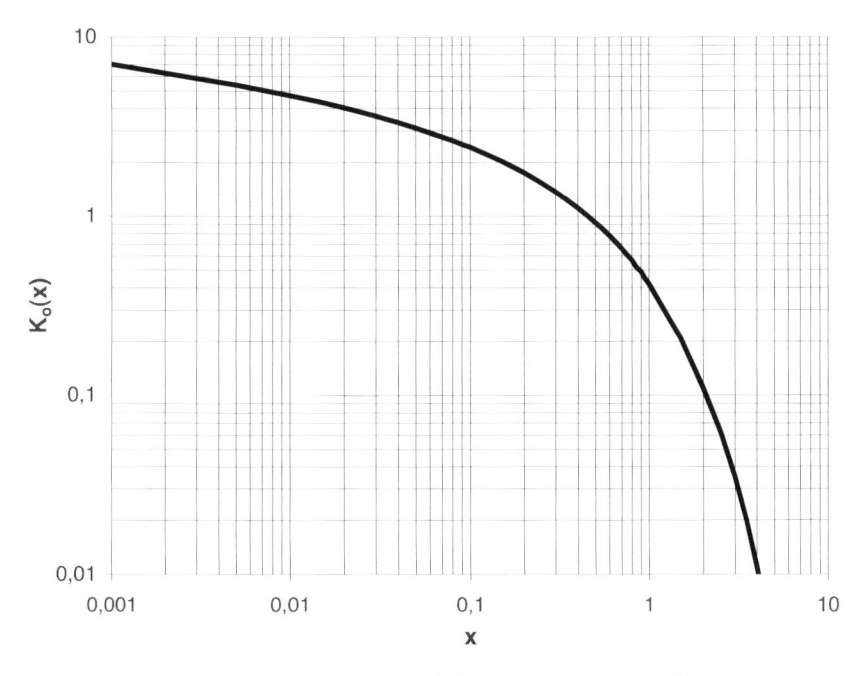

*Figure 6.10 Graph for determination of $\lambda$.*

where time is in minutes. Matching the Theis curve to late data gives us homogeneous aquifer parameters using the equations

$$T_f = \frac{Q}{4\pi s_{MP}} \qquad (6.26)$$

$$S_f + S_m = \frac{240 T_f t_{MP}}{r^2} \qquad (6.27)$$

The two matchings should produce the same value of $T_f$, i.e. the value of $s_{MP}$ must be the same. $\lambda$ is determined from Equation 6.23. Bourdet and Gringarten showed that the effect of dual porosity in monitoring wells appears only within a certain distance from the pumped well. For $\lambda > 1.78$ monitoring well data will show only homogeneous aquifer behavior and therefore data can be used to determine only $T_f$ and $S_f + S_m$.

Example 6.2 Determination of $\lambda$

From a pumping test, we have the following data: $Q = 20$ m³/h and $T_f = 0.01$ m²/s. The horizontal section the curve has a drawdown value of 0.1 m. From Equation 6.23 we get

$$K_0(\sqrt{\lambda}) = \frac{2\pi T_f s}{Q} = \frac{2\pi \cdot 0.01 \cdot 0.1 \cdot 3600}{20} = 1.13$$

From Figure 6.10 we determine $\sqrt{\lambda} = 0.4$ or $\lambda = 0.16$.

## 6.5 Streltsova

*Streltsova (1976)* solved the equation for drawdown in monitoring wells in a dual-porosity aquifer and published tabular values for type curves for $\omega = 0.1$, 0.01 and 0.001. The mathematics applies the usual physics with parameters describing flow in fractures and matrix. If one of the $\omega$ values fits data, type curves provide for parameter evaluation directly. If not, curve matching must follow the procedure described in the previous chapter. Type-curves for $\omega = 0.1$ are shown in Figure 6.11, for $\omega = 0.01$ in Figure 6.12 and for $\omega = 0.001$ in Figure 6.13.

Parameter evaluation follows the type-curve procedure. From the early Match Point we get

$$T_f = \frac{Q}{4\pi s_{MP}} \quad \text{and} \quad S_f = \frac{240 T_f t_{MP}}{r^2}$$

Matching late data gives

$$T_f = \frac{Q}{4\pi s_{MP}} \quad \text{and} \quad S_f + S_m = \frac{240 T_f t_{MP}}{r^2}$$

Streltsova defined the parameter $r/B$ as $r/B = r\sqrt{\dfrac{2K_m}{T_f h_m}}$, where $h_m$ is the matrix thickness in the layer model or the block length in the cubic model. From this we get

$$K_m = \frac{(r/B)^2 T_f h_m}{2r^2},$$

where the value of $r/B$ is taken from the type curve providing the best data fit.

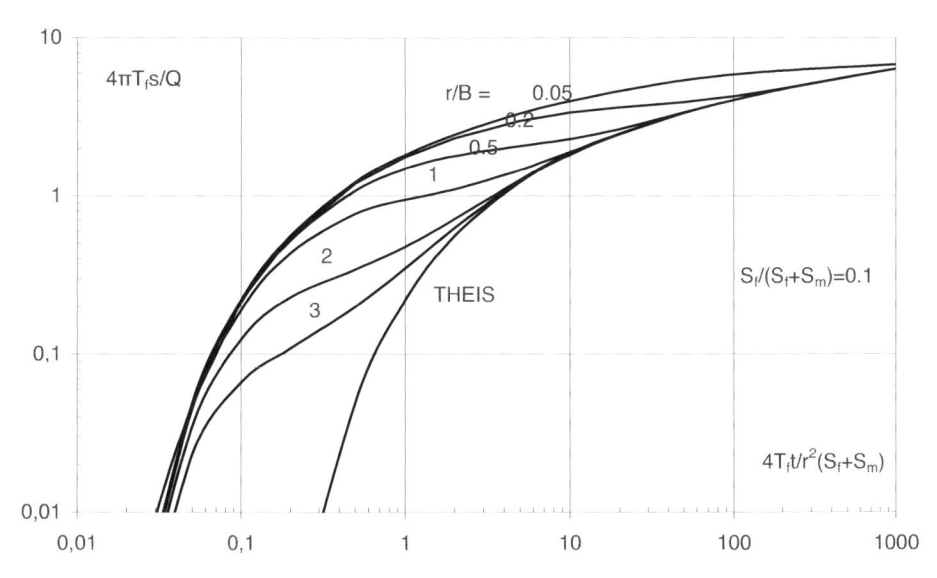

*Figure 6.11 Dimensionless drawdown in a monitoring well in a dual-porosity aquifer for $\omega = 0.1$, (Streltsova, 1976).*

It is interesting to note that Streltsova's type-curves are also valid for a leaky artesian sand aquifer. The curves are the same, but the physical meaning of the parameters is different. $T_f$ is now the transmissivity in the pumped aquifer, $S_f$ would be the confined storage coefficient and $S_f + S_m$ represent the late

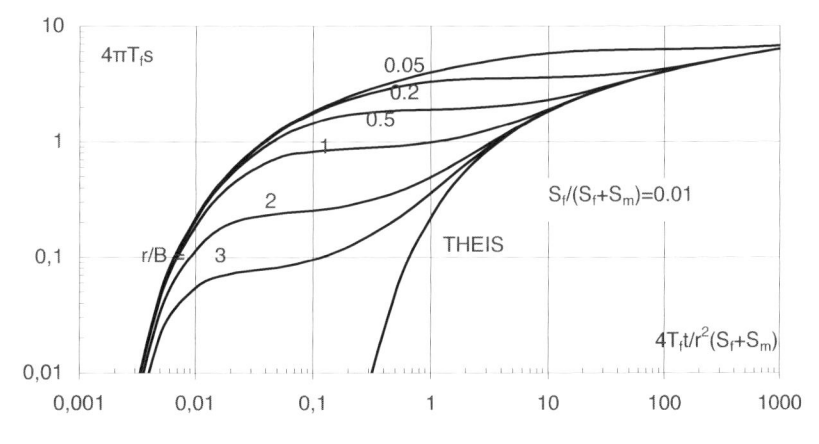

*Figure 6.12 Dimensionless drawdown in a monitoring well in a dual-porosity aquifer for $\omega = 0.01$, (Streltsova, 1976).*

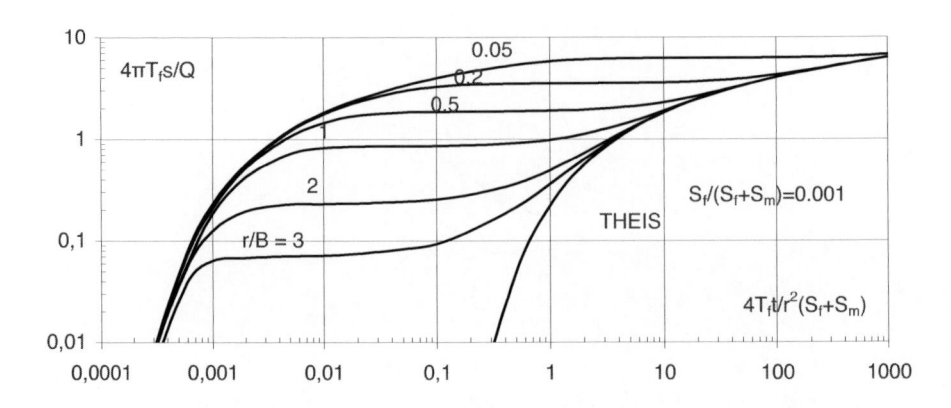

*Figure 6.13 Dimensionless drawdown in a monitoring well in a dual-porosity aquifer for $\omega = 0.001$, (Streltsova, 1976).*

storage coefficient, when the drawdown has reached the top of the confining layer. $h_m$ is the thickness of the confining layer, $K_m$ is the hydraulic conductivity of the confining layer and $S_m$ is the storage coefficient of the water-table aquifer.

In addition, *Streltsova (1978)* gave the solution for the drawdown $s_m$ in matrix. The matrix has a smaller drawdown than the fractures and there is a time delay compared to the drawdown in fractures. The drawdown in matrix again depends on the value of $\omega$. Figure 6.14 shows dimensionless drawdown in matrix as a function of dimensionless time for $\omega = 0.1$. Figures 6.15 and 6.16 show the drawdown curves in matrix for $\omega = 0.01$ and $\omega = 0.001$ respectively.

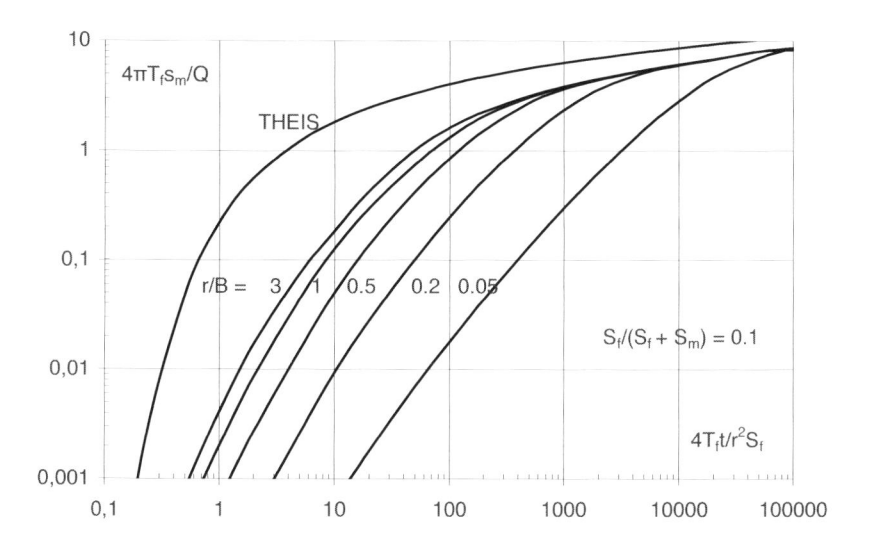

*Figure 6.14. Drawdown in matrix for* $\omega = 0.1$.

The Theis curve shown is the drawdown in fractures. Note the use of $S_f$ to define dimensionless time. By applying the type-curve method we can determine $T_f$ and $S_f$ in the usual way and $S_f + S_m$ is found from the relationship $S_f + S_m = S_f / \omega$. The parameter $B$ is

$$B = \frac{2}{\pi}\sqrt{\frac{T_f h_m^2}{T_m}}$$

This gives the matrix transmissivity

$$T_m = \frac{4(r/B)^2 T_f h_m^2}{\pi^2 r^2}.$$

Figures 6.14 to 6.16 are also valid for a leaky two-aquifer system. The drawdown on top of the confining clay layer, i.e. in the upper aquifer, now replaces the drawdown in matrix.

# FRACTURED AQUIFERS

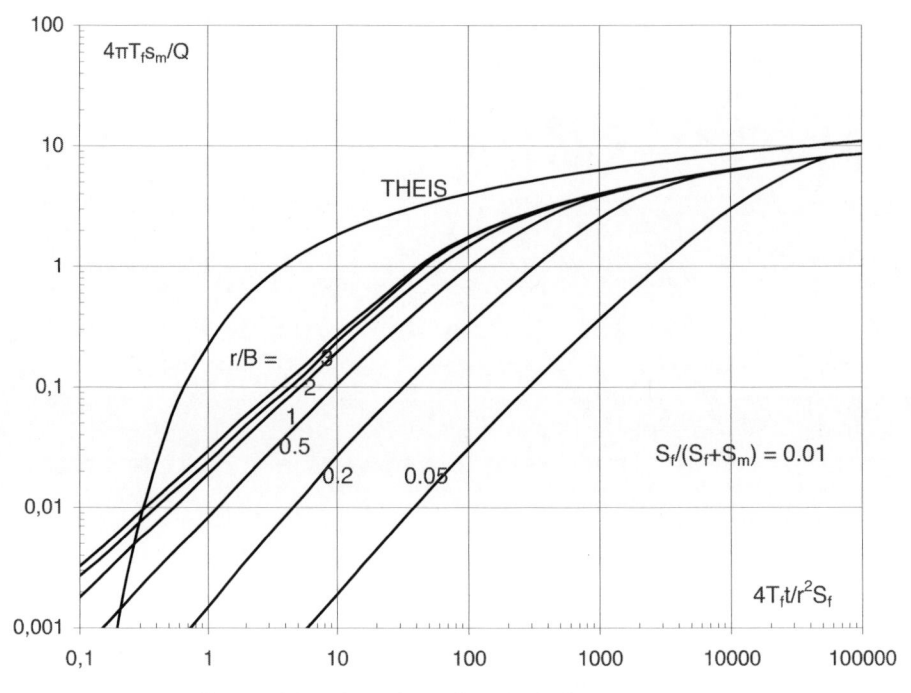

*Figure 6.15. Drawdown in matrix for $\omega = 0.01$.*

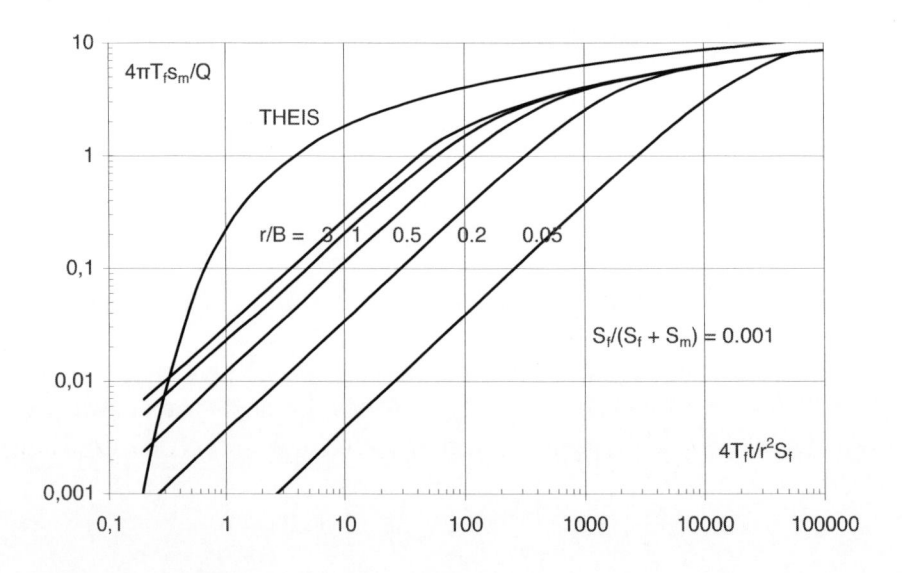

*Figure 6.16. Drawdown in matrix for $\omega = 0.001$.*

## Example 6.3

Figure 6.17 shows drawdown data from a monitoring well during the test from Example 6.1. The table shows selected data from the test, while the figures display the full data set.

The distance to the monitoring well is 18 m. Drawdown data match Streltsova's type curves for $\omega = 0.1$ and $r/B = 0.5$. The Match Point co-ordinates are $(s_{MP}, t_{MP}) = (0.02$ m, 1.8 min$)$ on the data sheet for late data and $(s_D, t_D) = (1,1)$ on the type-curve sheet. We can now calculate the hydraulic parameters as follows:

| t (min) | s (cm) | t (min) | s (cm) | t (min) | s (cm) | t (min) | s (cm) | t (min) | s (cm) |
|---|---|---|---|---|---|---|---|---|---|
| 0.33 | 0.9 | 2.83 | 3.5 | 10 | 4.3 | 46 | 5.8 | 140 | 7.1 |
| 0.5 | 0.8 | 3 | 3.2 | 11 | 4.4 | 50 | 5.2 | 150 | 7.2 |
| 0.67 | 1.9 | 3.5 | 3.6 | 12 | 4.2 | 54 | 5.8 | 170 | 8 |
| 0.83 | 2.3 | 4 | 3.7 | 14 | 4.2 | 58 | 5.9 | 190 | 7.8 |
| 1 | 1.8 | 4.5 | 3.6 | 16 | 4.5 | 62 | 6 | 200 | 8.3 |
| 1.17 | 2.5 | 5 | 4.1 | 18 | 4.4 | 66 | 6 | 220 | 8.4 |
| 1.33 | 2.7 | 5.5 | 3.4 | 20 | 4.4 | 70 | 6.2 | 240 | 8.8 |
| 1.5 | 2.6 | 6 | 3.8 | 22 | 4.6 | 75 | 6 | 260 | 8.5 |
| 1.67 | 2.8 | 6.5 | 4.2 | 24 | 5.2 | 80 | 6.5 | 280 | 8.4 |
| 1.83 | 2.6 | 7 | 3.8 | 26 | 4.6 | 85 | 6.2 | 300 | 9.4 |
| 2 | 2.6 | 7.5 | 3.4 | 28 | 5.2 | 90 | 6.5 | 320 | 9.6 |
| 2.17 | 3.1 | 8 | 3.5 | 30 | 4.8 | 100 | 6.5 | 340 | 9.8 |
| 2.33 | 3.3 | 8.5 | 3.9 | 34 | 5.3 | 110 | 6.8 | 359.5 | 9.6 |
| 2.5 | 3.28 | 9 | 4.2 | 38 | 5 | 120 | 7.1 | | |
| 2.67 | 3.4 | 9.5 | 4.1 | 42 | 5.6 | 130 | 6.8 | | |

$$T_f = \frac{Q}{4\pi s_{MP}} = \frac{15.1}{4\pi \cdot 0.02 \cdot 3600} = 0.0167 \text{ m}^2/\text{s}$$

$$S_f + S_m = \frac{240 T_f t_{MP}}{r^2} = \frac{240 \cdot 0.0167 \cdot 1.8}{18^2} = 0.0223$$

$$S_f = \omega(S + S_m) = 0.1 \cdot 0.0223 = 0.00223$$

10 % of the ground water flows in fractures. If we choose $h_m = 1$m, the hydraulic conductivity of matrix becomes

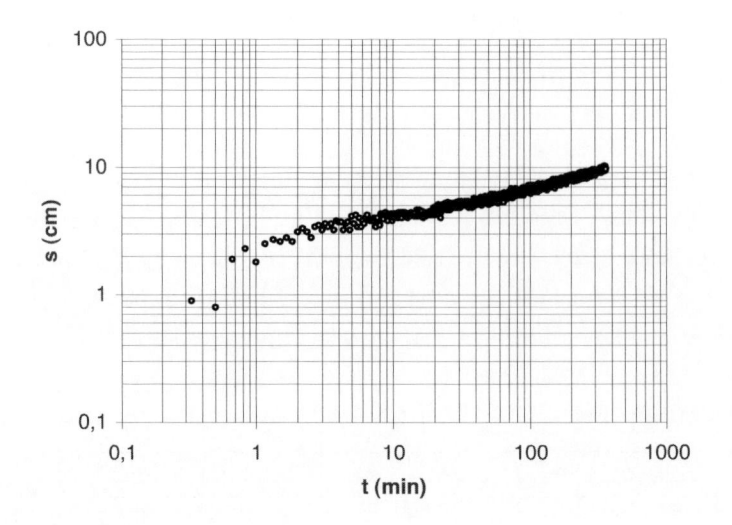

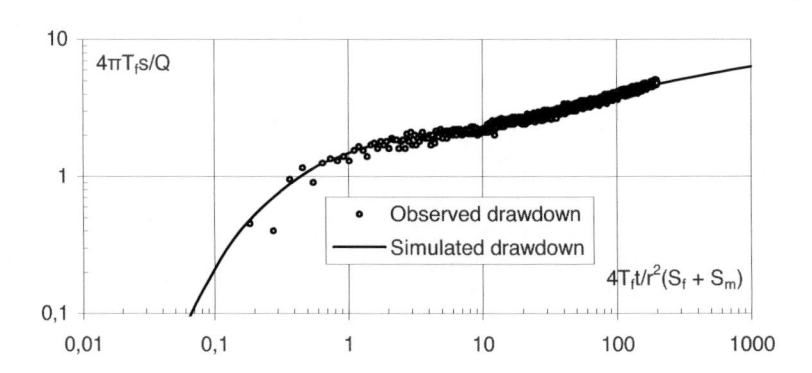

*Figure 6.17. Drawdown data (top) in a monitoring well from the test in Example 6.1 and simulated drawdown in a dimensionless plot (bottom).*

$$K_m = \frac{(r/B)^2 T_f h_m}{2r^2} = \frac{0.5^2 \cdot 0.0167 \cdot 1}{2 \cdot 18^2} = 6.4 \cdot 10^{-6} \text{ m/s}.$$

With these hydraulic parameters, Figure 6.17 shows the simulated drawdown in the monitoring well.

## Example 6.4

This example illustrates the use of Streltsova's type-curves in a leaky sand aquifer system at Esbjerg, Denmark *(Geological Survey of Denmark, 1971)*. A clay layer separates the lower pumped aquifer from the upper aquifer. Monitoring Well No. 580 is screened in the pumped aquifer and Well No. 737 is screened in the upper aquifer. The distances to the wells are 607 m and 561 m, respectively. The discharge rate is 130.4 m³/h. The thickness of the clay layer is 18 m. The tables show drawdown data from the pumped well and from the two monitoring wells. Figure 6.18a and b show the data plots. Changes in atmospheric pressure influence late data.

Pumped well

| t (min) | sw (m) | t (min) | sw (m) | t (min) | sw (m) |
|---|---|---|---|---|---|
| 2 | 1.525 | 35 | 1.745 | 720 | 1.894 |
| 3 | 1.555 | 40 | 1.761 | 960 | 1.923 |
| 4 | 1.58 | 50 | 1.762 | 1200 | 1.943 |
| 5 | 1.595 | 60 | 1.768 | 1320 | 1.95 |
| 6 | 1.615 | 75 | 1.78 | 1680 | 1.98 |
| 7 | 1.626 | 90 | 1.785 | 2100 | 1.996 |
| 8 | 1.637 | 120 | 1.794 | 2560 | 2.005 |
| 9 | 1.6458 | 150 | 1.8 | 3000 | 2.05 |
| 10 | 1.656 | 180 | 1.805 | 4000 | 2.055 |
| 12 | 1.67 | 210 | 1.81 | 4500 | 2.055 |
| 14 | 1.686 | 240 | 1.82 | 5400 | 2.055 |
| 16 | 1.7 | 300 | 1.835 | 6000 | 2.055 |
| 18 | 1.71 | 360 | 1.845 | 6800 | 2.05 |
| 20 | 1.71 | 420 | 1.855 | 7400 | 2.12 |
| 25 | 1.732 | 480 | 1.858 | 8500 | 2.076 |
| 30 | 1.735 | 600 | 1.875 | 9000 | 2.055 |

In this example, we use the following notation: In the pumped aquifer, $T$ and $S$, in the upper aquifer, $T'$ and $S'$, in the clay layer separating the aquifers, $K'$ and thickness $m'$, and therefore, $\omega = S/(S+S')$. Drawdown data from well No. 580 matches Streltsova's type curves for a pumped aquifer with $\omega = 0.1$, (Figure 6.11) and data from Well No. 737 matches Streltsova's type curves for the upper aquifer (Figure 6.14).

Data from the pumped well has a close resemblance to that found in dual-porosity aquifers.

# FRACTURED AQUIFERS

| Well No. 580 |   |   |   | Well No. 737 |   |   |   |
|---|---|---|---|---|---|---|---|
| t/r² (min/m²) | s (m) | t/r² (min/m²) | s (m) | t/r² (min/m²) | s (m) | t/r² (min/m²) | s (m) |
| 0.00022 | 0.0435 | 0.016 | 0.2 | 0.00024 | 0.002 | 0.0095 | 0.13 |
| 0.00043 | 0.068 | 0.0192 | 0.22 | 0.00058 | 0.002 | 0.0108 | 0.13 |
| 0.00058 | 0.08 | 0.022 | 0.23 | 0.00065 | 0.003 | 0.013 | 0.125 |
| 0.0011 | 0.095 | 0.026 | 0.24 | 0.00075 | 0.007 | 0.016 | 0.12 |
| 0.0013 | 0.102 | 0.032 | 0.28 | 0.00095 | 0.011 | 0.0185 | 0.95 |
| 0.0017 | 0.11 | 0.0393 | 0.29 | 0.00105 | 0.015 | 0.022 | 0.135 |
| 0.0021 | 0.12 | 0.045 | 0.3 | 0.0014 | 0.012 | 0.024 | 0,14 |
| 0.0025 | 0.122 | 0.055 | 0.31 | 0.0018 | 0.027 | 0.025 | 0.13 |
| 0.0034 | 0.128 | 0.065 | 0.33 | 0.00217 | 0.024 |   |   |
| 0.00435 | 0.135 | 0.079 | 0.34 | 0.0025 | 0.037 |   |   |
| 0.0052 | 0.14 | 0.095 | 0.38 | 0.003 | 0.05 |   |   |
| 0.00695 | 0.155 | 0.12 | 0.38 | 0.0036 | 0.068 |   |   |
| 0.0085 | 0.165 | 0.15 | 0.383 | 0.0044 | 0.079 |   |   |
| 0.0094 | 0.17 | 0.18 | 0.38 | 0.005 | 0.09 |   |   |
| 0.011 | 0.175 | 0.2 | 0.39 | 0,0061 | 0.096 |   |   |
| 0.0125 | 0.18 | 0.22 | 0.42 | 0.0072 | 0.11 |   |   |
| 0.014 | 0.19 | 0.24 | 0.39 | 0.0085 | 0.118 |   |   |

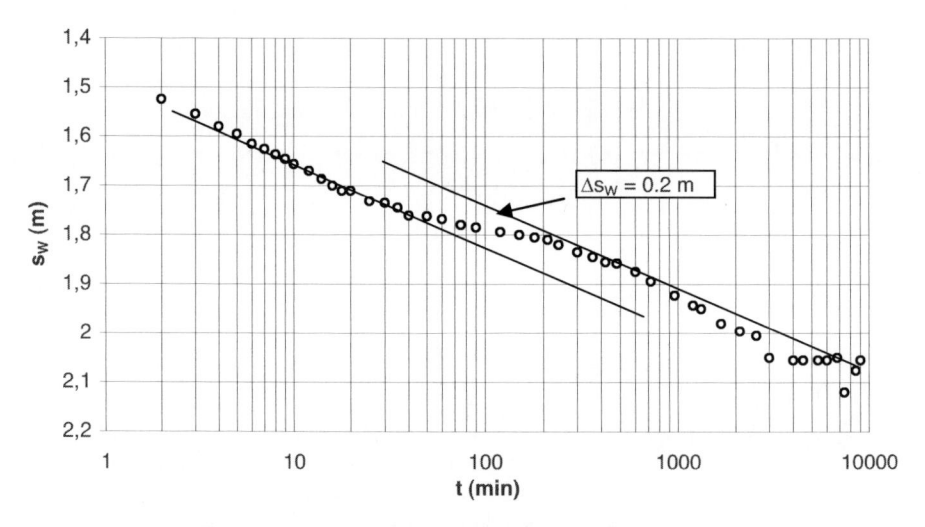

*Figure 6.18a Data plot of drawdown from the pumped well.*

The transmissivity is

$$T = \frac{0.183 Q}{\Delta s_w} = \frac{0.183 \cdot 130.4}{0.2 \cdot 3600} = 0.033 \text{ m}^2/\text{s}$$

For the monitoring well in the pumped aquifer, Well No. 580, we get Match Point co-ordinates $(s_{MP}, t/r^2_{MP}) = (0.1, 0.0012)$ and the type-curve parameter $r/B = 1$. Hydraulic parameters are

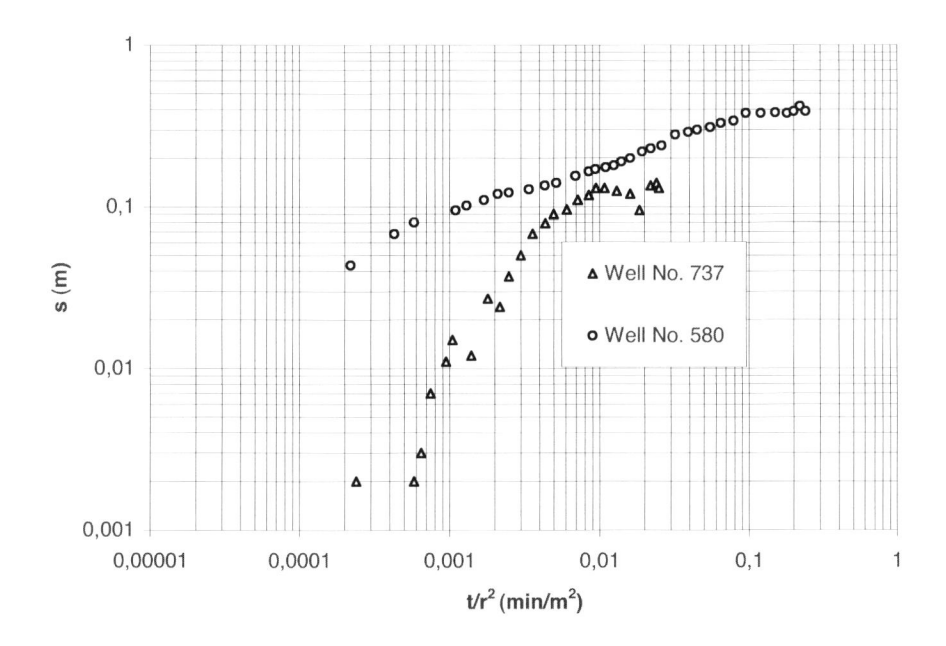

*Figure 6.18b Drawdown data from monitoring wells.*

$$T = \frac{Q}{4\pi s_{MP}} = \frac{130.4}{4\pi \cdot 0.1 \cdot 3600} = 0.0288 \text{ m}^2/\text{s}$$

$$S + S' = 240 T t_{MP}/r^2 = 240 \cdot 0.0288 \cdot 0.0012 = 0.0083$$

$$S = \omega(S + S') = 0.1 \cdot 0.0083 = 0.00083$$

$$K' = \frac{(r/B)^2 T m'}{2r^2} = \frac{1^2 \cdot 0.0288 \cdot 18}{2 \cdot 607^2} = 7.03 \cdot 10^{-7} \text{ m/s}$$

When we match the type curve, Figure 6.14, with data from the monitoring well in the upper aquifer, we have to use the same value of $s_{MP}$ as before, and use 0.1 times the value of $t/r^2$ to get $t/r^2{}_{MP}$. Therefore, we get the same value of $T, S$ and $S'$ and the value $r/B = 2$. We can now calculate the transmissivity of the upper aquifer as

$$T' = \frac{4(r/B)^2 Tm'^2}{\pi^2 r^2} = \frac{4 \cdot 2^2 \cdot 0.0288 \cdot 18^2}{\pi^2 \cdot 561^2} = 5.4 \cdot 10^{-5} \text{ m}^2/\text{s}$$

Figure 6.19 shows the simulated drawdowns using the calculated values of hydraulic parameters.

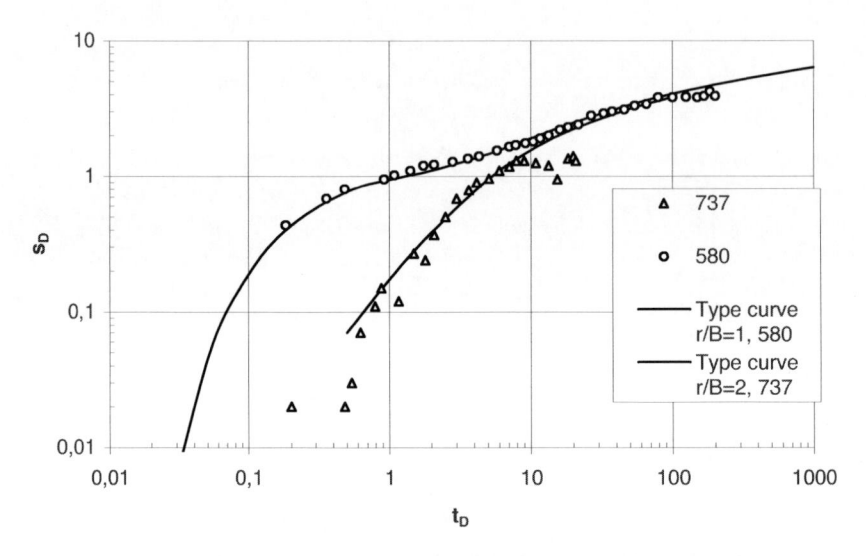

*Figure 6.19 Simulated drawdown from Example 6.4 in a dimensionless plot.*

## 6.6 Najurieta

*Najurieta (1980)* wrote an excellent paper on the interpretation of well test data from fractured aquifers with dual-porosity using ground water notation and with parallels to oil field notation. Because storage changes from a small value in fractures to a large bulk value, he defined a time-dependent storage function through a time-dependent diffusivity. Najurieta's mathematics is therefore different from earlier works but based on the classical dual-porosity concept. As we shall see later, his method in some cases provides an excellent data match where other methods fail. Najurieta made a model for blocks as well as layers. However, there seems to be a consensus that the stratum case is most common, *(Sorensen, 1981)*. Figure 6.6 shows a schematic of the stratum case aquifer.

6.6.1 The stratum case

The fracture diffusivity is

$$\eta_f = \frac{K_f h_f}{S_f} = \frac{T_f}{S_f}$$

and for matrix

$$\eta_m = \frac{K_m h_m}{S_m} = \frac{T_m}{S_m}.$$

Najurieta's general solution to the flow problem in a fractured aquifer built of layers reads

$$s = \frac{Q}{4\pi T_f} W\left(\frac{r^2}{4\eta_{CO} t}\right) \tag{6.28}$$

We recognize equation 6.28 as Theis's equation with a modified diffusivity $\eta_{CO}$ (composite) combining storage in fractures and matrix:

$$\eta_{CO} = \frac{T_f}{S_f + S_m \sqrt{\frac{t}{\tau}} \tanh\sqrt{\frac{\tau}{t}}}, \text{ where} \tag{6.29}$$

$$\tau = \frac{h_m^2}{4\gamma \eta_m} \tag{6.30}$$

Equation 6.28 is the exact solution when $t \gg \tau$, and a good approximation when $\tau > t > r^2/0.05\eta_{CO}$. We note that Equation 6.28 is valid for the pumped well and for monitoring wells. For the pumped well we can obtain the Jacob-Cooper equation as

$$s_w = \frac{0.183 Q}{T_f} \log \frac{135 t \eta_{CO}}{r_w^2} \tag{6.31}$$

Equation 6.29 may seem unusual but as we shall see, it describes the sought behavior of storage.

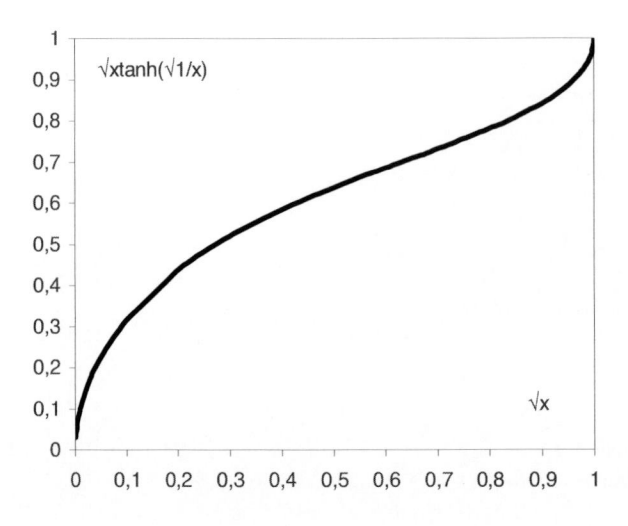

*Figure 6.20 The function* $\sqrt{x}\tanh\left(1/\sqrt{x}\right)$.

For small times (see Figure 6.20) $\sqrt{\dfrac{t}{\tau}}\tanh\sqrt{\dfrac{\tau}{t}} \cong 0$, and Equation 6.29 reduces to $\eta_{COS} = \eta_f = \dfrac{T_f}{S_f}$. For large times, $\sqrt{\dfrac{t}{\tau}}\tanh\sqrt{\dfrac{\tau}{t}} \cong 1$, and Equation 6.29 becomes $\eta_{COL} = \dfrac{T_f}{S_f + S_m}$, i.e. Warren and Root's classical model is the basis for Najurieta's model. In a semi-log plot, we obtain an early straight line for fracture behavior and a late straight line for homogeneous aquifer behavior.

When $t > \tau$, data points fall on the late straight line. This gives the useful information that the transition curve and the late straight line coincide at the time $t \cong \tau$.

Figure 6.21 shows theoretical drawdown curves (Equation 6.31) for the following parameters: $T = 0.05$ m²/s, $S_f = 0.000110$, $S_m = 0.01$, $\tau = 200$ min, $r_w = 0.1$ m.

The drawdown curves in Figure 6.21 show the well-known dual-porosity behavior. For large distances, the dual-porosity behavior vanishes and we see only the homogeneous aquifer behavior representing bulk storage (broken lines). Najurieta's model does not allow pseudo-steady state to appear because the transition phase is non-stationary. The change in slope from the transition curve to the late straight line may erroneously be interpreted as the influence from a barrier boundary.

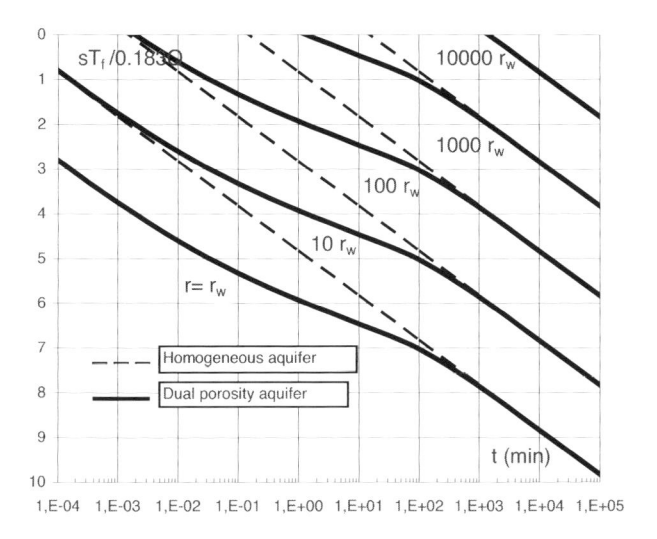

*Figure 6.21 Theoretical drawdown curves from Najurieta's model.*

Figure 6.21 also indicates that in the pumped well, only part of the transition curve, or in some cases, no part at all may appear. The time-dependency of the storage term appears from Figure 6.20. To see the exact variation with time and to prepare type curves for monitoring well drawdown, some calculations are necessary, (Sorensen, 1981).

Since $\eta_{CO} = \dfrac{T_f}{S_{CO}}$ we get

$$u_{CO} = \frac{r^2 S_{CO}}{4T_f t} = \frac{r^2 (S_f + S_m)}{4T_f t} \cdot \frac{S_{CO}}{S_f + S_m} = u_{COL} \frac{S_{CO}}{S_f + S_m} \tag{6.32}$$

From Equation 6.29 we obtain

$$\frac{S_{CO}}{S_f + S_m} = \frac{S_f + S_m \sqrt{\dfrac{t}{\tau}} \tanh \sqrt{\dfrac{\tau}{t}}}{S_f + S_m} = \frac{\varphi}{1+\varphi} + \frac{1}{1+\varphi} \sqrt{\dfrac{t}{\tau}} \tanh \sqrt{\dfrac{\tau}{t}} \tag{6.33}$$

where $\varphi = \dfrac{S_f}{S_m}$ has been introduced. $\dfrac{t}{\tau}$ is rewritten using Equation 6.30,

$$\frac{\tau}{t} = \frac{h_m^2}{4\gamma\eta_m t} = \frac{h_m^2 S_m}{4\gamma K_m h_m t} = \frac{h_m^2 S_m}{4\gamma K_m h_m t} \cdot \frac{S_f + S_m}{S_f + S_m} \cdot \frac{T_f}{T_f} \cdot \frac{r^2}{r^2} = \psi \cdot u_{COL}$$

(6.34)

where 
$$\psi = \frac{h_m^2 S_m T_f}{\gamma K_m h_m r^2 (S_f + S_m)}$$
(6.35)

Now we may write Equation 6.33 as

$$\frac{S_{CO}}{S_f + S_m} = \frac{\varphi}{1+\varphi} + \frac{1}{1+\varphi}\sqrt{\frac{1}{\psi u_{COL}}} \tanh\sqrt{\psi u_{COL}}$$
(6.36)

The general drawdown equation now becomes

$$s = \frac{Q}{4\pi T_f} W(u_{CO}), \text{ where}$$
(6.37)

$$u_{CO} = u_{COL}\left(\frac{\varphi}{1+\varphi} + \frac{1}{1+\varphi}\sqrt{\frac{1}{\psi u_{COL}}} \tanh\sqrt{\psi u_{COL}}\right)$$
(6.38)

Figure 6.22 illustrates the change with time of storage (Equation 6.36). We see the transition from fracture storage for small times to bulk storage for large times.

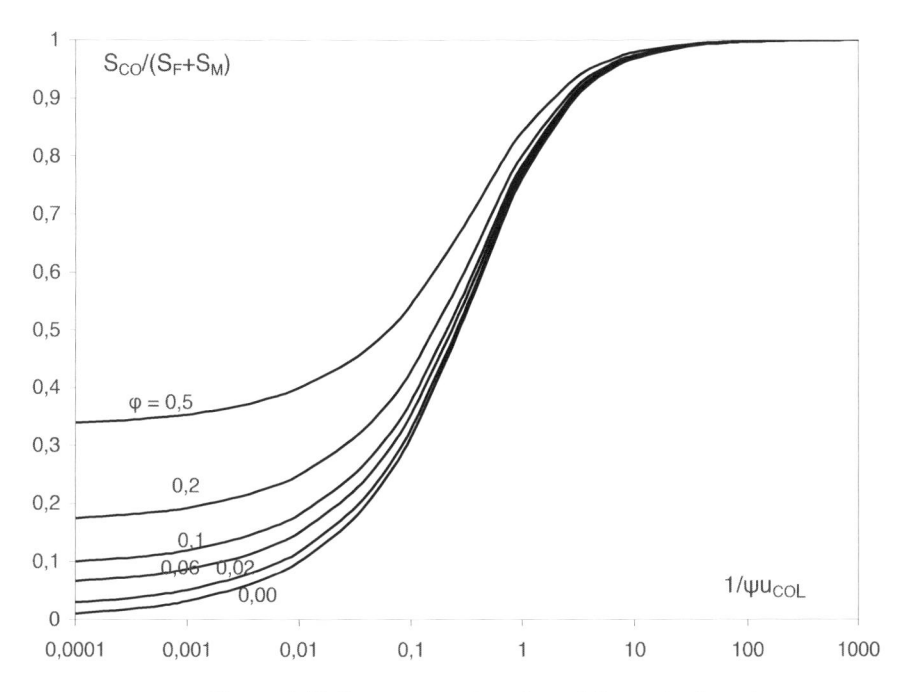

*Figure 6.22 Storage as a function of dimensionless time.*

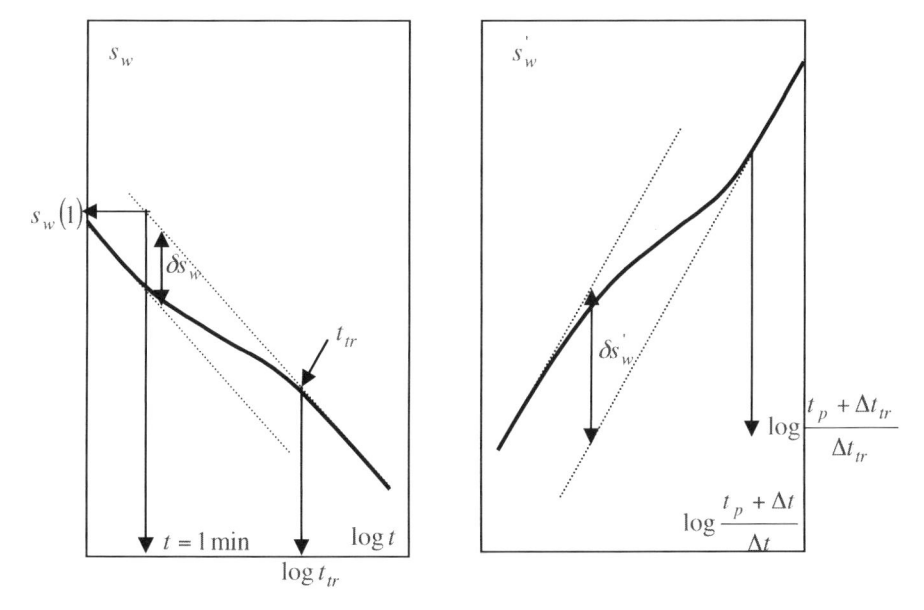

*Figure 6.23 Interpretation of drawdown data (left) and recovery data (right), after Najurieta (1980).*

## 6.6.2 Evaluation of hydraulic parameters for the pumped well

*Drawdown data*

Normally data from the pumped well can only be used to determine fracture transmissivity and assess skin loss and well performance indicators. Determination of storage requires that skin loss and turbulent loss are negligible. Using residual drawdown data, however, eliminates the skin loss problem, but this data cannot give storage values. In spite of this, Najurieta devised a method for calculating hydraulic parameters, which is analogous to the method of Warren and Root.

First, we use late data to calculate $T_f$ and $S_f + S_m$ from a semi-log plot:

$$T_f = \frac{0.183Q}{\Delta s_w} \qquad (6.39)$$

We extrapolate the late straight line to $t = 1$ min. (Figure 6.23) and note the drawdown $s_w(1)$. From Equation 6.31 we get

$$\eta_{COL} = \frac{r_w^2}{135} 10^{s_w(1)T_f/0.183Q} \text{, where} \qquad (6.40)$$

$$\eta_{COL} = \frac{T_f}{S_f + S_m}.$$

The vertical displacement between the two lines is $\delta s_w$ and we get as before $\omega = \frac{S_f}{S_f + S_m} = 10^{-\delta s_w/\Delta s_w}$. From this, we can determine $S_f$:

$$S_f = \frac{T_f}{\eta_{COL}} 10^{-\delta s_w/\Delta s} = \frac{\omega T_f}{\eta_{COL}} \qquad (6.41)$$

Similarly, we can determine $S_m$ since $S_m = S_f\left(\frac{1}{\omega} - 1\right)$ and inserting Equation 6.41:

$$S_m = \frac{T_f}{\eta_{COL}}(1 - \omega) \qquad (6.42)$$

Finally an approximate value of $\tau$ is taken as the time where the transition curve meets the late straight line, i.e. $\tau \cong t_{tr}$ (transition time), see Figure 6.23.

The procedure above is essentially the same as using the intersection between the straight lines and the zero drawdown line and then determining $t_0$. For the pumped well the latter method is often impractical but it is always possible to read the drawdown after 1 min.

If the early straight line does not develop it is only possible to determine $T_f$ and $S_f + S_m$.

*Recovery data*

The equation for the residual drawdown is, Figure 4.5 and Equation 6.31

$$s'_w = \frac{0.183Q}{T_f}\left(\log\frac{135(t_p+\Delta t)\eta_{CO}(t_p+\Delta t)}{r_w^2} - \log\frac{135\Delta t\, \eta_{CO}(\Delta t)}{r_w^2}\right) = \frac{0.183Q}{T_f}\log\left(\frac{\eta_{CO}(t_p+\Delta t)}{\eta_{CO}(\Delta t)}\cdot\frac{t_p+\Delta t}{\Delta t}\right) \quad (6.43)$$

For large values of $\Delta t$ we get $\eta_{CO}(\Delta t) = \eta_{CO}(t_p+\Delta t) = \eta_{COL}$, and Equation 6.43 reduces to

$$s'_w = \frac{0.183Q}{T_f}\log\frac{t_p+\Delta t}{\Delta t} \quad (6.44)$$

Equation 6.44 represents the late straight line in Figure 6.23. From this equation, we can determine $T_f$ as usual. For small values of $\Delta t$ we get $\eta(\Delta t) = \eta_f$ and Equation 6.43 becomes

$$s'_w = \frac{0.183Q}{T_f}\log\frac{\eta_{COL}}{\eta_f}\frac{t_p+\Delta t}{\Delta t} \quad (6.45)$$

From this equation, we can calculate $\dfrac{\eta_{COL}}{\eta_f} = \dfrac{S_f}{S_f + S_m} = \omega$.

Normally the pumping time is long enough to produce the late straight line where $\eta_{CO}(t_p) \cong \eta_{COL}$. Therefore, we can measure the vertical displacement between the two lines in the recovery plot, Figure 6.23, and calculate

$$\omega = 10^{-\delta s'_w / \Delta s'_w}.$$

Recovery data cannot be used to calculate $S_f$ and $S_m$. $\tau$ determined in the usual way is $\tau \cong \Delta t_{tr}$.

Skin effect calculations are the same as those described in Chapter 4, but in the equations, we should replace $T/S$ by $\eta_{CO}$. Figure 6.24 shows the significance of $\tau$ and $S_f / S_m$ for the recovery curve. The larger $\tau$ is, the later data falls on the late straight line. The significance of $S_f / S_m$ is analogous to the significance of $\omega$.

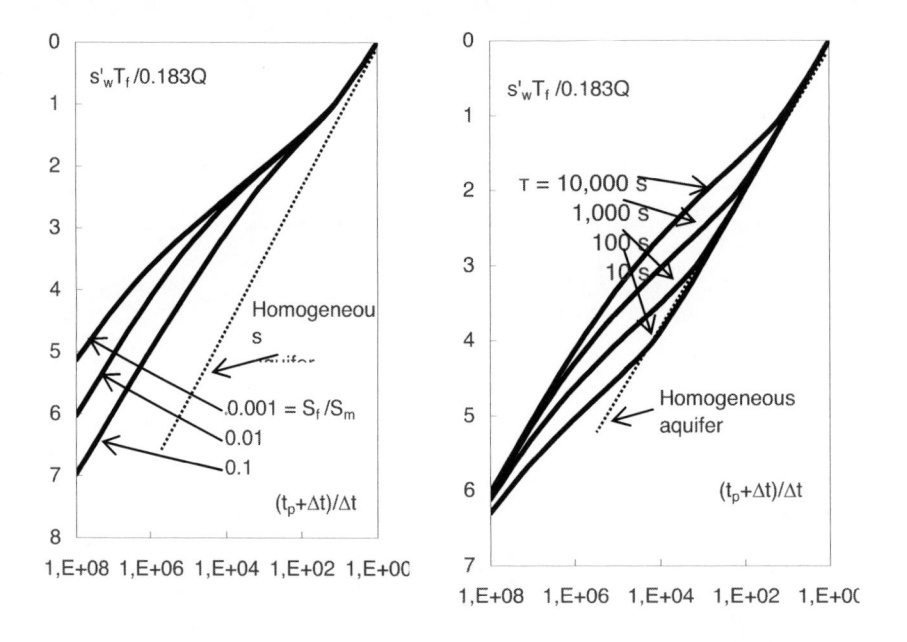

*Figure 6.24 The variation of $S_f / S_m$ for $\tau = 10,000\,s$ (left) and $\tau$ (right).*

DUAL POROSITY

Example 6.4 Najurieta's method for the pumped well

Data for Example 6.4 is from the recovery in an oil well from *Najurieta (1980)*, Figure 6.25. Basic data is $Q = 0.0108$ m³/s; the drawdown at the end of the test is $s_w = 293.7$ m, $t_p = 516{,}667$ min. and $r_w = 0.114$ m. The table shows recovery data.

| Δt (min) | (t_p+Δt)/Δt | s'_w (m) |
|---|---|---|
| 1 | 516668 | 103.364 |
| 2 | 258335 | 98.026 |
| 4 | 129168 | 93.682 |
| 8 | 64584 | 91.313 |
| 16 | 32293 | 86.377 |
| 32 | 16147 | 82.922 |
| 64 | 8073 | 80.750 |
| 128 | 4037 | 74.926 |
| 256 | 2019 | 70.187 |
| 512 | 1010 | 62.290 |
| 1024 | 505.6 | 57.157 |
| 2048 | 253.3 | 52.419 |

Fracture transmissivity determined from Figure 6.25 is:

$$T_f = \frac{0.183 Q}{\Delta s_w'} = \frac{0.183 \cdot 0.0108}{21.7} = 9.1 \cdot 10^{-5} \text{ m}^2/\text{s}.$$

The test has not shown the early straight line, and well bore storage influences the first 3 – 4 data points. To determine $\omega$ it is necessary to read the value of $\frac{t_p + \Delta t}{\Delta t}$ at the transition point between the transition curve and the late straight line and from this determine $\tau$. After this, we estimate $\delta s_w'$. After some trials, we obtain a good fit with $\frac{t_p + \Delta t}{\Delta t} = 4100$, from which, using $\Delta t = \Delta t_{tr} \cong \tau$, we get $\tau = 126$ min. and $\delta s_w' = 72.9$ m. From the relation

$$\omega = \frac{S_f}{S_f + S_m} = \frac{1}{1 + \frac{S_m}{S_f}} = 10^{-\delta s_w' / \Delta s_w'} \text{ we determine}$$

$$\frac{S_f}{S_m} = 0.000437$$

71

# FRACTURED AQUIFERS

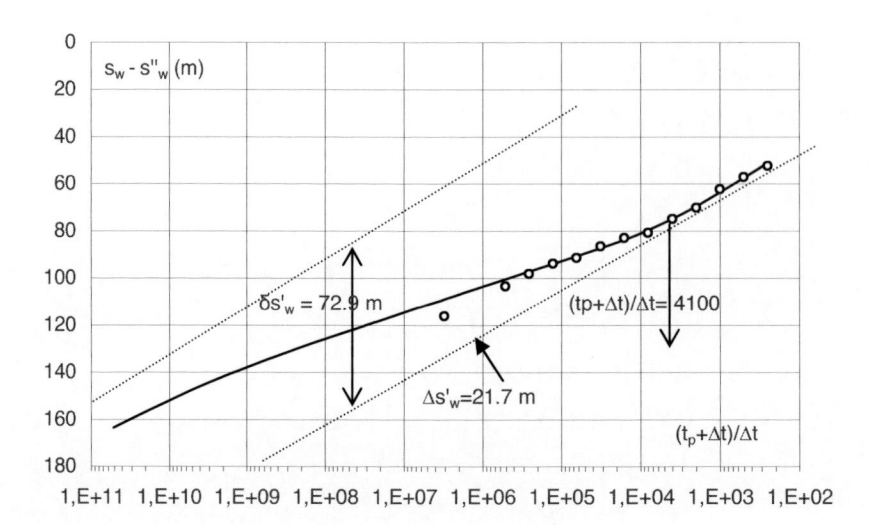

*Figure 6.25 Recovery data from an oil well.*

We can check the interpretation by writing down the equation for the residual drawdown from Equation 6.43:

$$s_w - s_w'' = \frac{0.183Q}{T_f} \log\left(\frac{\eta_{CO}(t_p + \Delta t)}{\eta_{CO}(\Delta t)} \cdot \frac{t_p + \Delta t}{\Delta t}\right) =$$

$$\frac{0.183Q}{T_f} \log\left(\frac{\varphi + \sqrt{\frac{\Delta t}{\tau}} \tanh\sqrt{\frac{\tau}{\Delta t}}}{\varphi + \sqrt{\frac{t_p + \Delta t}{\tau}} \tanh\sqrt{\frac{\tau}{t_p + \Delta t}}} \cdot \frac{t_p + \Delta t}{\Delta t}\right) =$$

$$21.7 \log\left(\frac{0.000437 + \sqrt{\frac{\Delta t}{126}} \tanh\sqrt{\frac{126}{\Delta t}}}{0.000437 + \sqrt{\frac{516667 + \Delta t}{126}} \tanh\sqrt{\frac{126}{516667 + \Delta t}}} \cdot \frac{516667 + \Delta t}{\Delta t}\right)$$

This equation is the full line in Figure 6.25. $\omega$ becomes

$$\omega = 10^{-72.9/21.7} = 0.000437.$$

This shows that 0.0437 % of the oil flows in fractures. In the oil industry, laboratory results often allow further analyses. In this case, drill core samples have given $h_m = 5.2$ m, and therefore from Equation 6.30 we can obtain matrix diffusivity:

$$\eta_m = \frac{h_m^2}{4\gamma\tau} = \frac{5.2^2}{4 \cdot 1.78 \cdot 126 \cdot 60} = 5 \cdot 10^{-4} \text{ m}^2/\text{s}.$$

In addition, from core samples, the specific matrix storage is $S_{sm} = 2.5 \cdot 10^{-5}$ m$^{-1}$ and we can calculate matrix hydraulic conductivity:

$$K_m = \eta_m S_{sm} = 5 \cdot 10^{-4} \cdot 2.5 \cdot 10^{-5} = 1.25 \cdot 10^{-8} \text{ m/s}.$$

The skin calculation requires additional data: $S_m = 0.0131$, $S_f = 5.85 \cdot 10^{-6}$, and from monitoring data, we get for $\Delta t = 16$ min.: $\dfrac{t_p + \Delta t}{\Delta t} = 32{,}292$ and the residual drawdown $s_w - s_w'' = 94$ m. The skin factor, found from Equation (4.17) since $t_p \gg \Delta t$, is:

$$\sigma = 1.1513\left(\frac{s_w - s_w'}{\Delta s} - \log\frac{\eta_{CO}(16)}{r_w^2} - 2.13\right), \text{ where}$$

$$\eta_{CO}(16) = \frac{9.1 \cdot 10^{-5}}{5.86 \cdot 10^{-6} + 0.0131\sqrt{\dfrac{16}{126}}\tanh\sqrt{\dfrac{126}{16}}} = 0.0196 \text{ m}^2/\text{s}$$

The skin factor becomes

$$\sigma = 1.1513\left(\frac{293.4 - 94}{21.7} - \log\frac{0.0196}{0.114^2} - 2.13\right) = 7.93$$

The skin loss is

$$s_s = \frac{\sigma Q}{2\pi T_f} = \frac{7.93 \cdot 0.0108}{2\pi \cdot 0.000091} = 150 \text{ m}.$$

The effective radius is

$$r_{ef} = r_w e^{-\sigma} = 0.114 e^{-7.93} = 0.000041 \text{ m}.$$

The well efficiency is

$$V = \frac{s_w - s_s}{s_w} = \frac{293.7 - 150}{293.7} = 0.49, \text{ and}$$

Clogging index and clogging factor become

$$C_i = 1/V = 2.04 \text{ and } C_f = 1 - V = 0.51$$

### 6.6.3 Drawdown in the aquifer

Najurieta's general solution 6.28 is valid for monitoring wells. Type-curves are based on Equations 6.37 and 6.38 where we have two parameters, $\varphi$ and $\psi$. Figure 6.26 shows stratum case type curves for $\varphi = \dfrac{S_f}{S_m} = 0$ and different values of

$$\psi = \frac{h_m^2 S_m T_f}{\gamma K_m h_m r^2 (S_f + S_m)}.$$

The family of type curves in the figure all approach the Theis curve for large times where homogeneous aquifer behavior prevails. With constant hydraulic aquifer parameters, $\psi$ is determined by the distance to the monitoring well. The larger the distance, the closer the curve is to the Theis curve. Again, it is possible to erroneously match data with Theis curves alone and accept the effect of barrier boundaries. The matching is most accurate if the test is long enough to define the Theis curve.

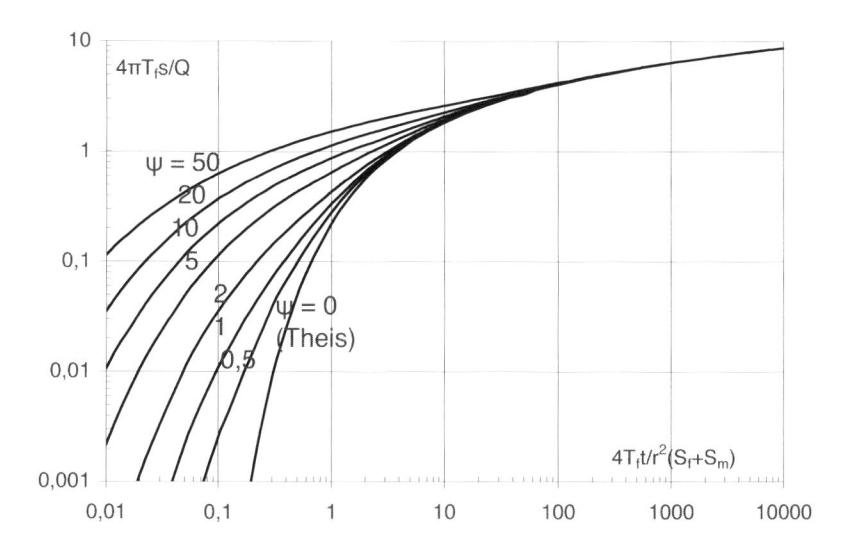

*Figure 6.26 Type curves for drawdown in the aquifer according to Najurieta's stratum model.*

The significance of the parameter $\varphi$ is illustrated in Figure 6.27 where $\psi = 1$. By comparing with the curves in Figure 6.26, it appears difficult to decide which values of $\varphi$ and $\psi$ to choose. Determining these parameters may not be unique. However, fracture transmissivity and bulk storage are easily determined if the Theis curve is seen. The method is the type-curve matching procedure:

$$T_f = \frac{Q}{4\pi s_{MP}} \qquad S_f + S_m = \frac{240 T_f t_{MP}}{r^2}.$$

As mentioned above it is difficult to determine $\omega$ from monitoring well data, but if we know $\omega$ from the pumped well data, then from Equation 6.35:

$$\psi = \frac{h_m^2 S_m T_f}{\gamma K_m h_m r^2 (S_f + S_m)} = \tau \cdot \frac{4 T_f \omega}{r^2 S_f} \qquad (6.46)$$

$\tau$ can be determined from data and $S_f$ calculated and after that, also $S_m$.

FRACTURED AQUIFERS

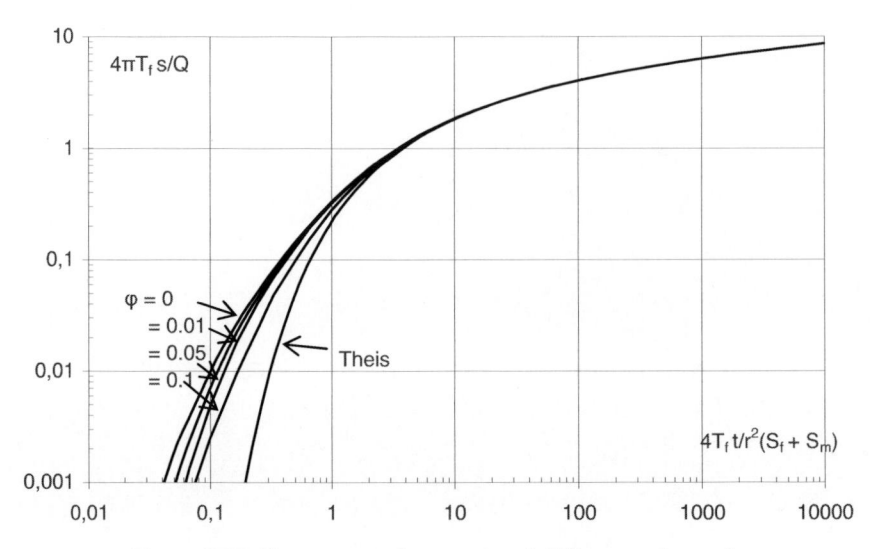

*Figure 6.27. Type-curves for $\psi = 1$ and different values of $\varphi$.*

Example 6.5. Najurieta's method for monitoring wells, stratum case

Data is from a monitoring well in a highly stratified and horizontally fractured chalk aquifer in central Zealand, Denmark, *(Geological Survey of Denmark, 1976)*. The discharge rate was 76.5 m³/h.

| t (min) | s (m) | t (min) | s (m) | t (min) | s (m) | t (min) | s (m) | t (min) | s (m) |
|---|---|---|---|---|---|---|---|---|---|
| 3 | 0.001 | 25 | 0.02 | 210 | 0.118 | 1080 | 0.28 | 4800 | 0.44 |
| 5 | 0.002 | 30 | 0.023 | 240 | 0.13 | 1200 | 0.285 | 5200 | 0.41 |
| 6 | 0.003 | 35 | 0.029 | 270 | 0.135 | 1380 | 0.295 | 5700 | 0.42 |
| 7 | 0.004 | 40 | 0.033 | 300 | 0.15 | 1620 | 0.31 | 6200 | 0.49 |
| 8 | 0.004 | 50 | 0.041 | 360 | 0.16 | 1740 | 0.32 | 6980 | 0.49 |
| 9 | 0.005 | 60 | 0.047 | 420 | 0.17 | 1860 | 0.33 | 7700 | 0.51 |
| 10 | 0.006 | 75 | 0.053 | 480 | 0.18 | 2220 | 0.34 | 8600 | 0.49 |
| 12 | 0.007 | 90 | 0.06 | 540 | 0.195 | 2580 | 0.38 | 10000 | 0.52 |
| 14 | 0.009 | 105 | 0.0685 | 600 | 0.21 | 2940 | 0.4 | 12000 | 0.55 |
| 16 | 0.0115 | 120 | 0.077 | 720 | 0.23 | 3300 | 0.41 | | |
| 18 | 0.013 | 150 | 0.089 | 840 | 0.25 | 3660 | 0.425 | | |
| 20 | 0.016 | 180 | 0.1 | 960 | 0.27 | 4020 | 0.41 | | |

The distance to the monitoring well is $r = 1213$ m. The table shows drawdown data and Figure 6.28 the data plot.

Matching with Najurieta's type-curves for $\phi = 0$ gives the Match Point values $(s_{MP}, t_{MP}) = (0.13 \text{ m}, 79 \text{ min})$ and $\psi = 2$. We get

$$T_f = \frac{Q}{4\pi s_{MP}} = \frac{76.5}{4\pi \cdot 0.13 \cdot 3600} = 0.013 \text{ m}^2/\text{s}$$

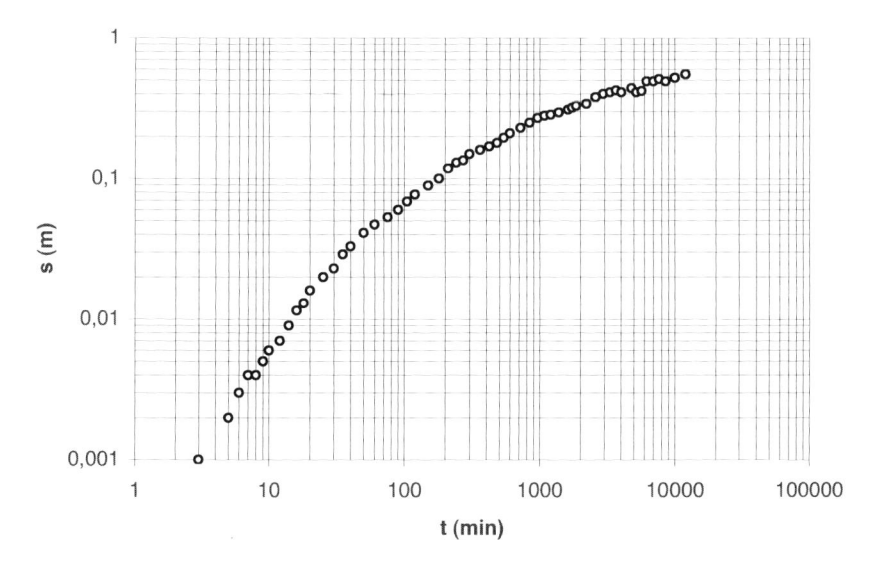

*Figure 6.28 Drawdown data from a monitoring well in a fractured chalk aquifer.*

$$S_f + S_m = \frac{240 T_f t_{MP}}{r^2} = \frac{240 \cdot 0.013 \cdot 79}{1213^2} = 0.000168$$

Figure 6.29 shows the simulated drawdown in a dimensionless plot using the values above. From the pumped well we have $\omega = 0.01$ and therefore, $S_f = \omega(S_f + S_m) = 0.01 \cdot 0.000168 = 0.00000168$. From Figure 6.29 we get the dimensionless transition time 5 and therefore

$$5 = \frac{240 T_f \tau}{r^2 (S_f + S_m)} \Rightarrow \tau = 5 \cdot 1213^2 \cdot 0.000168 / 240 \cdot 0.013 = 396 \text{ min}$$

From Equation 6.46 we can get another estimate of $S_f$:

$$S_f = \tau \cdot \frac{4T_f \omega}{r^2 \cdot 2} = 396 \frac{4 \cdot 0.01 \cdot 0.01 \cdot 60}{1213^2 \cdot 2} = 0.00000323$$

We can now calculate matrix diffusivity:

$$\eta_m = \frac{h_m^2}{4\gamma\tau} = \frac{0{,}2^2}{4 \cdot 1.78 \cdot 396 \cdot 60} = 2.36 \cdot 10^{-7} \text{ m}^2/\text{s, where we have}$$

chosen the thickness of the matrix layer $h_m = 0.2$ m. We can now give an estimate of matrix transmissivity:

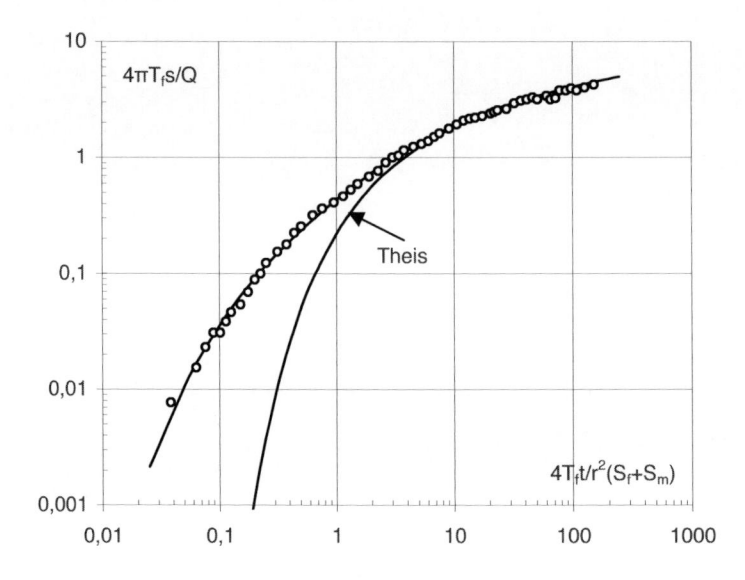

*Figure 6.29. Simulation of drawdown data from a monitoring well in a chalk aquifer, $\psi = 2$ .*

$$T_m = \eta_m S_m = 2.36 \cdot 10^{-7} \cdot (0.000168 - 0.00000168) = 3.93 \cdot 10^{-11} \text{ m}^2/\text{s}$$

or matrix hydraulic conductivity $K_m = 3.93 \cdot 10^{-11} / 0{,}2 = 1.96 \cdot 10^{-10}$ m/s.

### 6.6.4 The blocks case

For the blocks case, Najurieta gave the diffusivity

$$\eta_{CO} = \frac{T_f}{S_{CO}} = \frac{T_f}{S_f + S_m\left(\sqrt{\dfrac{t}{\tau}}\coth\sqrt{\dfrac{\tau}{t}} - \dfrac{t}{\tau}\right)} \qquad (6.47)$$

For small times, $\sqrt{\dfrac{t}{\tau}}\coth\sqrt{\dfrac{\tau}{t}} - \dfrac{t}{\tau} \cong 0$ and therefore, $\eta_{COS} = \dfrac{T_f}{S_f}$ as in the stratum case. For long times, $(t/\tau > 1)$ we have $\sqrt{\dfrac{t}{\tau}}\coth\sqrt{\dfrac{\tau}{t}} - \dfrac{t}{\tau} \cong \dfrac{1}{3}$ and therefore, $\eta_{COL} = \dfrac{3T_f}{3S_f + S_m}$.

For the pumped well we obtain as for the stratum case

$$s_w = \frac{0.183Q}{T_f}\log\frac{135t\eta_{CO}}{r_w^2} \qquad (6.48)$$

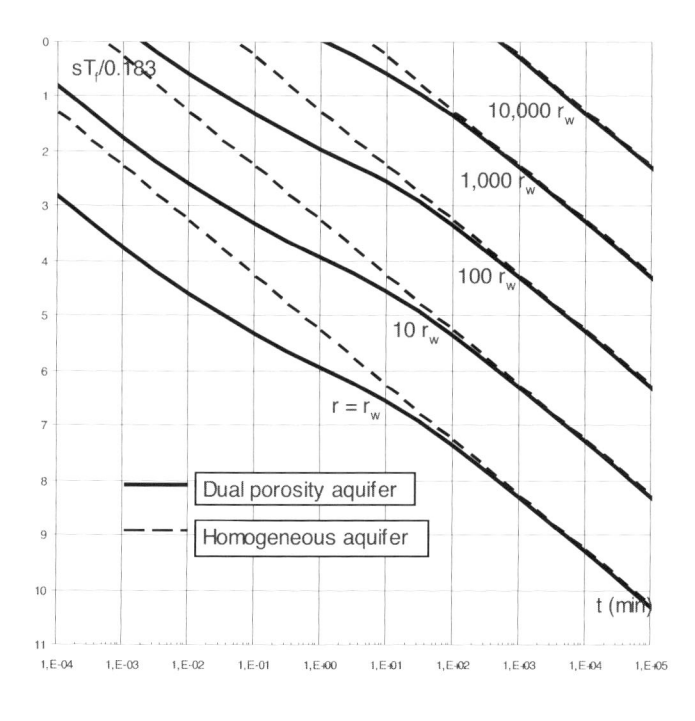

*Figure 6.30 Theoretical drawdown curves, Najurieta's blocks model.*

Figure 6.30 shows theoretical drawdown curves based on equation 6.48 for the following parameters: $T_f = 0.05 \, \text{m}^2/\text{s}$, $S_f = 0.0001$, $S_m = 0.01$, $\tau = 200$ min and $r_w = 0.1$ m.

For small times, the stratum (Figure 6.21) and block cases have the same drawdown (fractures predominate), but for long times, the drawdown is different because $\eta_{COL}$ is different.

For the general drawdown in the aquifer we get

$$u_{CO} = \frac{r^2 S_{CO}}{4T_f t} = \frac{r^2 (3S_f + S_m)}{3 \cdot 4T_f t} \cdot \frac{3S_{CO}}{3S_f + S_m} = u_{COL} \frac{3S_{CO}}{3S_f + S_m} \quad (6.49)$$

where $u_{COL} = \dfrac{r^2 (3S_f + S_m)}{12 T_f t}$.

From Equation 6.47 we obtain

$$\frac{3S_{CO}}{3S_f + S_m} = \frac{3\left(S_f + S_m \left(\sqrt{\dfrac{t}{\tau}} \coth \sqrt{\dfrac{\tau}{t}} - \dfrac{t}{\tau}\right)\right)}{3S_f + S_m} =$$

$$\frac{3\left(\varphi + \sqrt{\dfrac{t}{\tau}} \coth \sqrt{\dfrac{\tau}{t}} - \dfrac{t}{\tau}\right)}{1 + 3\varphi} \quad (6.50)$$

From Equation 6.30 we get

$$\frac{\tau}{t} = \frac{h_m^2}{4\gamma \eta_m t} = \frac{h_m^2 S_m}{4\gamma K_m h_m t} = \frac{h_m^2 S_m}{4\gamma K_m h_m t} \cdot \frac{3T_f}{3T_f} \cdot \frac{3S_f + S_m}{3S_f + S_m} \cdot \frac{r^2}{r^2} = \varepsilon u_{COL} \quad (6.51)$$

where

$$\varepsilon = \frac{3 h_m^2 S_m T_f}{\gamma K_m h_m r^2 (3S_f + S_m)} \quad (6.52)$$

Equation 6.50 now becomes

$$\frac{3S_{CO}}{3S_f + S_m} = \frac{3\left(\varphi + \sqrt{\frac{1}{\varepsilon u_{COL}}} \coth \sqrt{\varepsilon u_{COL}} - \frac{1}{\varepsilon u_{COL}}\right)}{1 + 3\varphi} \qquad (6.53)$$

The general drawdown equation for the blocks case becomes

$$s = \frac{Q}{4\pi T_f} W(u_{CO}) \qquad (6.54)$$

$$u_{CO} = u_{COL} \frac{3\left(\varphi + \sqrt{\frac{1}{\varepsilon u_{COL}}} \coth \sqrt{\varepsilon u_{COL}} - \frac{1}{\varepsilon u_{COL}}\right)}{1 + 3\varphi} \qquad (6.55)$$

Figure 6.31 shows the type-curves based on Equation 6.54.

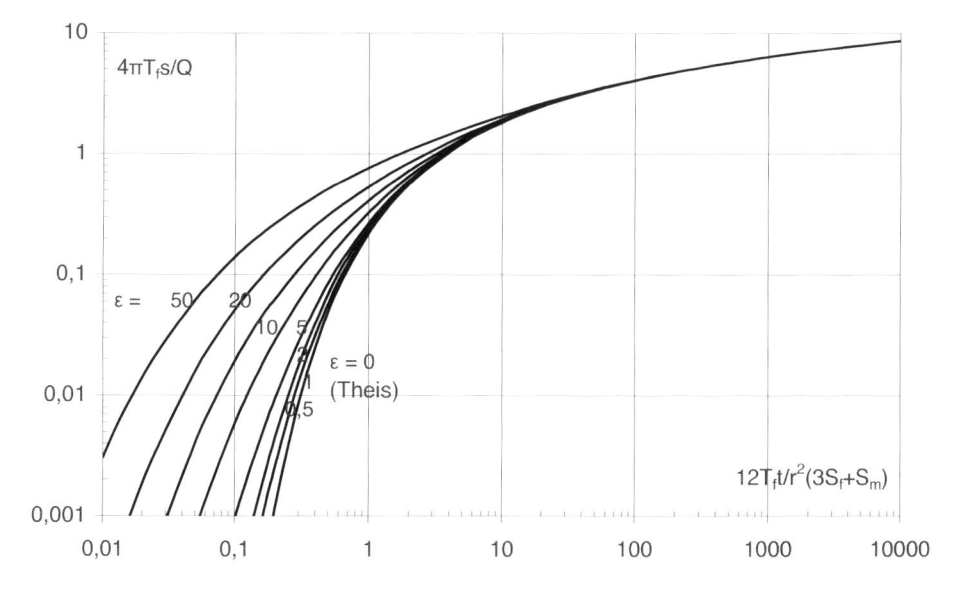

*Figure 6.31 Type-curves for drawdown in the aquifer according to Najurieta's blocks model with $\varphi = 0$.*

# FRACTURED AQUIFERS

The stratum case type-curves (Figure 6.26) are different from the blocks case type-curves because diffusivities are different. Both sets of type curves should produce the same value of fracture transmissivity, but storage coefficients will be different.

Example 6.6 Najurieta's method for monitoring wells, blocks case

We use the data from example 6.5. Matching with type curves for the block case gives the Match Point co-ordinates $(s_{MP}, t_{MP}) = (0.13\text{m}, 92 \text{ min})$. We get

$$T_f = \frac{Q}{4\pi s_{MP}} = \frac{76.5}{4\pi \cdot 0.13 \cdot 3600} = 0.013 \text{ m}^2/\text{s}$$

$$3S_f + S_m = \frac{12T_f t_{MP}}{r^2} = \frac{12 \cdot 0.013 \cdot 92 \cdot 60}{1213^2} = 0.000585$$

The blocks model gives the same fracture transmissivity as the stratum model, and higher matrix storage. Since $S_f \ll S_m$, we get roughly $S_m = 0.00058$.

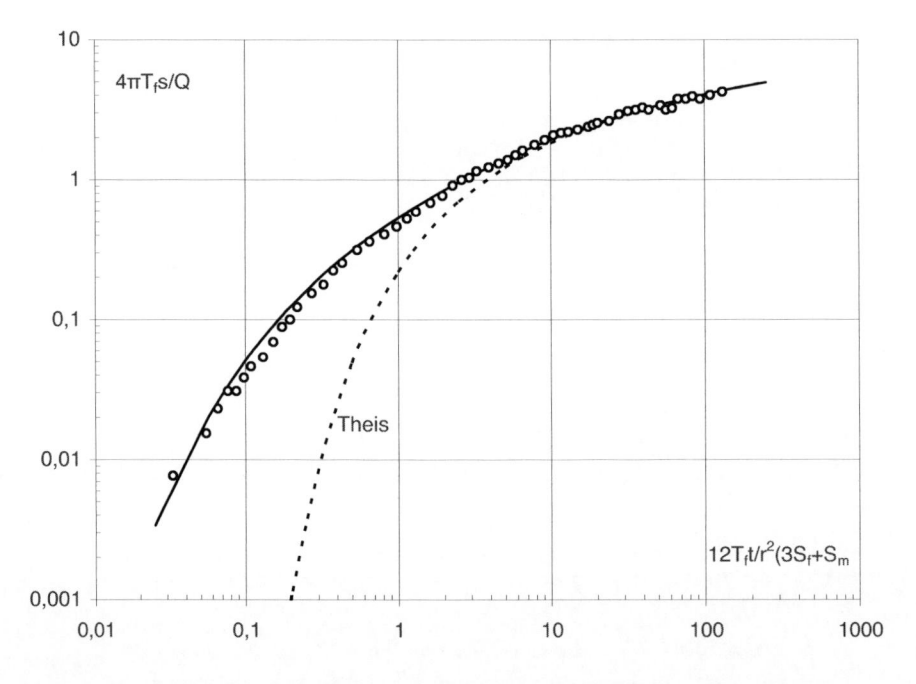

*Figure 6.32 Simulation of drawdown data from a monitoring well in a chalk aquifer, $\varepsilon = 20$.*

## 6.7 Estimating block size or layer thickness

In the dual-porosity models presented here, the block size or layer thickness and the matrix hydraulic conductivity always appear as a ratio. In Warren and Root's model, the parameter $\lambda$ defines the ratio, in Streltsova's model, the parameter $B$ does, and in Najurieta's model, there are two parameters, $\psi$ for layers and $\varepsilon$ for blocks. Therefore, to determine the block size, we need to know matrix hydraulic conductivity. These parameters are normally not available for water supply wells and therefore, block size or layer thickness cannot be determined.

*Braester (1984)* studied the influence of blocks size on the transition curve for a drawdown test in a naturally fractured reservoir. He concluded that for aquifers with ordered fractures and blocks, drawdown pressure changes are not sensitive enough to the variation in sizes of the blocks. A drawdown or recovery test can therefore not yield a solution unique for the size of the blocks.

*Braester and Zeitoun (1993)* used a stochastic model for flow through inhomogeneous, fractured dual-porosity aquifers. They assumed an ideal reservoir with a cylindrical inhomogeneity around the well and compared the solution for drawdown with Warren and Root's solution for an ideal homogeneous aquifer. The result was that homogeneous and inhomogeneous aquifers exhibit the same drawdown behavior, and drawdown curves could be fitted by different values of $\lambda$ and $\omega$ when attributed to an inhomogeneous or homogeneous aquifer. They concluded that the ambiguity in determining the two parameters makes the determination of them questionable.

The purpose of testing water supply wells is to obtain hydraulic parameters describing homogeneous aquifer behavior and use them for prediction of future drawdowns. The parameters $\lambda$, $B$, $\psi$ and $\varepsilon$ describe the transition from fracture storage to bulk storage and are of no interest for predictions. Therefore, the actual value of the parameters, including $\omega$, is not important. The transition period is, however, important to identify during test data interpretation because correct transition identification is a condition for a correct determination of homogeneous aquifer parameters.

# 7. Single vertical fracture

We now leave the uniformly fractured aquifers and move to aquifers in which the presence of a single fracture (vertical or horizontal) controls the ground water flow. This chapter is about vertical fractures.

The oil literature operates with three types of vertical fractures,

•   *The uniform flux fracture.* This type of fracture has a very large, but not infinitely large hydraulic conductivity. In other words, the fracture transmissivity is very large compared to the transmissivity in the surrounding rock formation. There is a uniform flux of water into the fracture per unit fracture area and there is a gradient along the fracture during pumping.

•   *The infinite hydraulic conductivity fracture.* The physical consequence of this assumption is that there is no gradient along the fracture. During pumping from this fracture, the drawdown is the same throughout the fracture. The drawdown increases with time.

•   *The finite hydraulic conductivity fracture.* In this case, the fracture has a hydraulic conductivity and an aperture.

*Russell and Truitt (1964)* and *van Everdingen and Meyer (1972)* published the first papers on vertical fractures. These works did not make it possible to analyze early data from well tests, and there were discrepancies between the papers and within the papers themselves *(Gringarten, Ramey and Raghaven, 1974)*. Gringarten, Ramey and Raghaven (1974) published the first comprehensive paper on the analyses of drawdown data from a well drilled in a vertical fracture. This paper is in strict oil-field notation, but Gringarten alone or together with others has published papers with analyzed cases, which are easier for ground water professionals to read. Apart from the 1974 paper the most important references are *Gringarten and Ramey (1973), Gringarten, Ramey and Raghaven (1975), Gringarten (1982), Gringarten and Witherspoon (1972), Raghaven, Uraiet and Thomas (1978), Cinco, Samaniego and Dominguez (1976 and 1978)* and *Cinco-Ley (1981)*.

The following account of interpretation methods is based primarily on the papers by Gringarten and Gringarten et al. *Gringarten et al (1974)* presented general equations for flow to a vertical fracture, where the uniform flux fracture and the infinite hydraulic conductivity fracture are special cases. Here, the general equations are not included, only equations which have a direct relevance to well test analysis.

FRACTURED AQUIFERS

## 7.1 The physical model

Figure 7.1 shows the vertical fracture set-up. The usual assumptions apply: the aquifer is confined and homogeneous, the fracture fully penetrates the aquifer, the well is drilled centrally in the fracture and screened over the full fracture. There are no turbulent losses and no well bore storage and the aquifer is infinite. The equations are valid also for a phreatic aquifer if the drawdowns are small compared the saturated aquifer thickness. The half-fracture length is $x_f$.

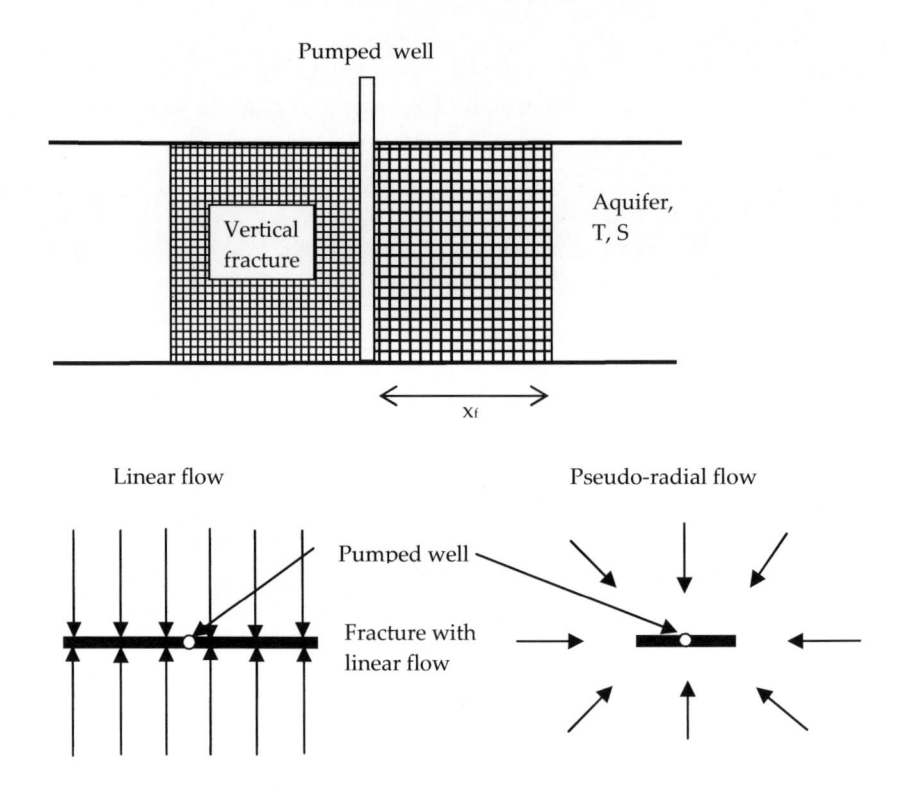

*Figure 7.1 A well in a vertical fracture showing fracture set-up (top), linear flow for small times (bottom left), and pseudo-radial flow for large times (bottom right). After Kruseman and de Ridder (2001).*

Water enters the well through the fracture only, and the fracture receives water from the aquifer. The aquifer transmissivity is $T$ and storage is $S$. The model does not consider fracture systems in the aquifer, but since we are dealing with fractured aquifers, the transmissivity is a fracture transmissivity. Storage is bulk storage, i.e. the aquifer behaves as the homogeneous aquifer.

86

## 7.2 The uniform flux vertical fracture

### 7.2.1 Drawdown in the pumped well

The dimensionless drawdown in the pumped well is *(Gringarten, Ramey and Raghaven, 1974)*,

$$s_{wD} = 2\sqrt{\pi t_D}\, erf\left(\frac{1}{2\sqrt{t_D}}\right) - Ei\left(-\frac{1}{4t_D}\right), \text{ where} \quad (7.1)$$

$$t_D = \frac{Tt}{x_f^2 S} \quad (7.2)$$

$$s_{wD} = \frac{4\pi T s_w}{Q}$$

$$-Ei(-z) = W(z), \text{ and}$$

$$erf(z) = \frac{2}{\sqrt{\pi}} \int_0^z e^{-x^2} dx, \text{ where}$$

$erf(z)$ is the error function. The error function has the following properties: $erf(0) = 0$ and $erf(\infty) = 1$. Note that the half-fracture length is used in defining dimensionless time, not the well radius. Written in aquifer parameters Equation 7.1 reads

$$s_w = \frac{Q}{4\pi T}\left(2\sqrt{\frac{\pi Tt}{x_f^2 S}}\, erf\left(\frac{1}{2}\sqrt{\frac{x_f^2 S}{Tt}}\right) + W\left(\frac{x_f^2 S}{4Tt}\right)\right) \quad (7.3)$$

For small times, flow in the aquifer is linear to the fracture, see Figure 7.1. We have that $W\left(\frac{x_f^2 S}{4Tt}\right) \cong 0$ and $erf\left(\frac{1}{2}\sqrt{\frac{x_f^2 S}{Tt}}\right) \cong 1$, and Equation 7.3 reduces to

$$s_w = \frac{Q}{2\sqrt{\pi T S x_f^2}}\sqrt{t} \quad (7.4)$$

For small times, drawdown increases with the square root of time and this is an important property that can be used to identify single fracture flow. As we shall see later, this is also the case for a single horizontal fracture.

For large times, the cone of depression is large compared to the fracture length and the flow becomes pseudo-radial, see Figure 7.1. We can obtain a simple equation for the drawdown. For small values of $z$ we have $\frac{1}{z}erf(z) \cong \frac{2}{\sqrt{\pi}}$. Thus, for large times the first member of the right hand side of Equation 7.1 becomes

$$2\sqrt{\pi t_D}\, erf\left(\frac{1}{2\sqrt{t_D}}\right) \cong \sqrt{\pi}\,\frac{2}{\sqrt{\pi}} = 2.$$

Also, for large times $W\left(\frac{1}{4t_D}\right) = -0.5772 + \ln(4t_D)$ and Equation 7.1 becomes

$$s_{wD} = 2 - 0.5772 + \ln(4t_D) = \ln(16.59 t_D) \tag{7.5}$$

Using the logarithm with base 10 and taking time in minutes, we get for the pseudo radial flow period

$$s_w = \frac{0.183Q}{T}\log\frac{996 Tt}{x_f^2 S} \tag{7.6}$$

We can use Equation 7.6 to calculate aquifer transmissivity by means of the straight-line method on semi-log paper. Note that 996 now replaces the number 135. Gringarten et al state that the semi-log straight line starts at approximately $t_D = 2$.

The drawdown develops three distinct time periods: An early linear flow period which in a log-log plot produces a straight line with slope 1:2, and a late period where the drawdown is log-log and follows the Theis equation (homogeneous aquifer behavior). Between these two periods, we have a transition curve.

The standard procedure for parameter evaluation is type-curve matching. Figure 7.2 shows Equation 7.1 in a log-log plot.

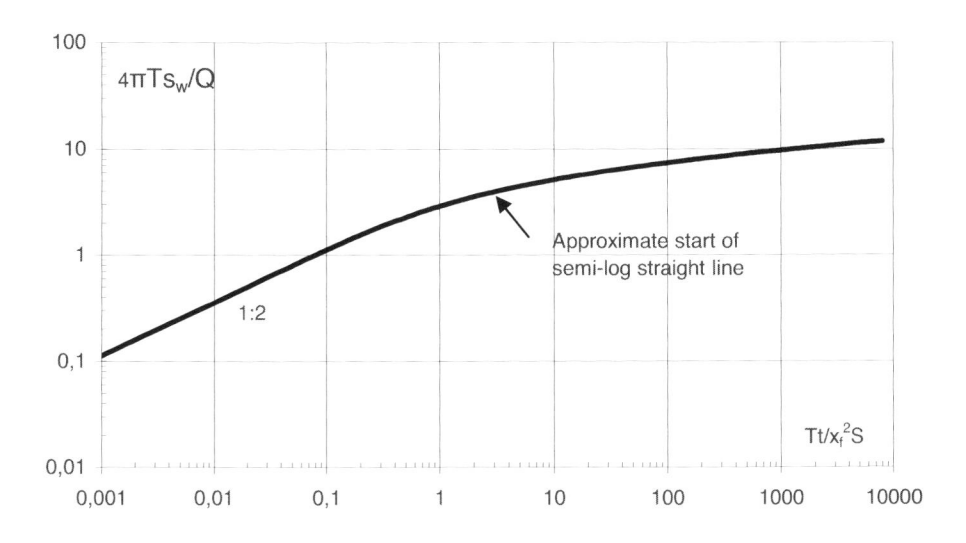

*Figure 7.2 Type curve for the drawdown in a well penetrating a uniform flux fracture. No well bore storage or turbulent losses.*

### 7.2.2 Drawdown in the aquifer

*Gringarten and Witherspoon (1972)* gave integral equations for the drawdown in monitoring wells. *Kruseman and de Ridder (2001)* published tabular values.

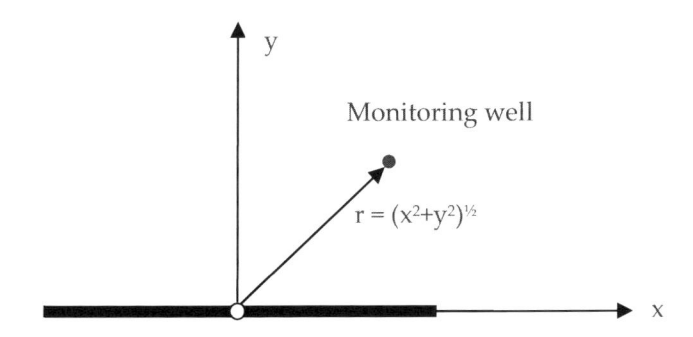

*Figure 7.3 Monitoring well near a vertical fracture.*

The following relationship defines the location of the monitoring well in relation to the fracture and the pumped well (Figure 7.3),

# FRACTURED AQUIFERS

$$r' = \frac{\sqrt{x^2 + y^2}}{x_f}$$

For a monitoring well drilled in the fracture or its extension ($r' = x/x_f$) the drawdown equation is

$$s = \frac{Q}{4\pi T} \cdot \frac{\sqrt{\pi}}{2} \int_0^{t_D} \left( erf\left(\frac{1-r'}{2\sqrt{\tau}}\right) + erf\left(\frac{1+r'}{2\sqrt{\tau}}\right) \right) \frac{d\tau}{\sqrt{\tau}} \qquad (7.7)$$

Figure 7.4 shows the type-curve plot.

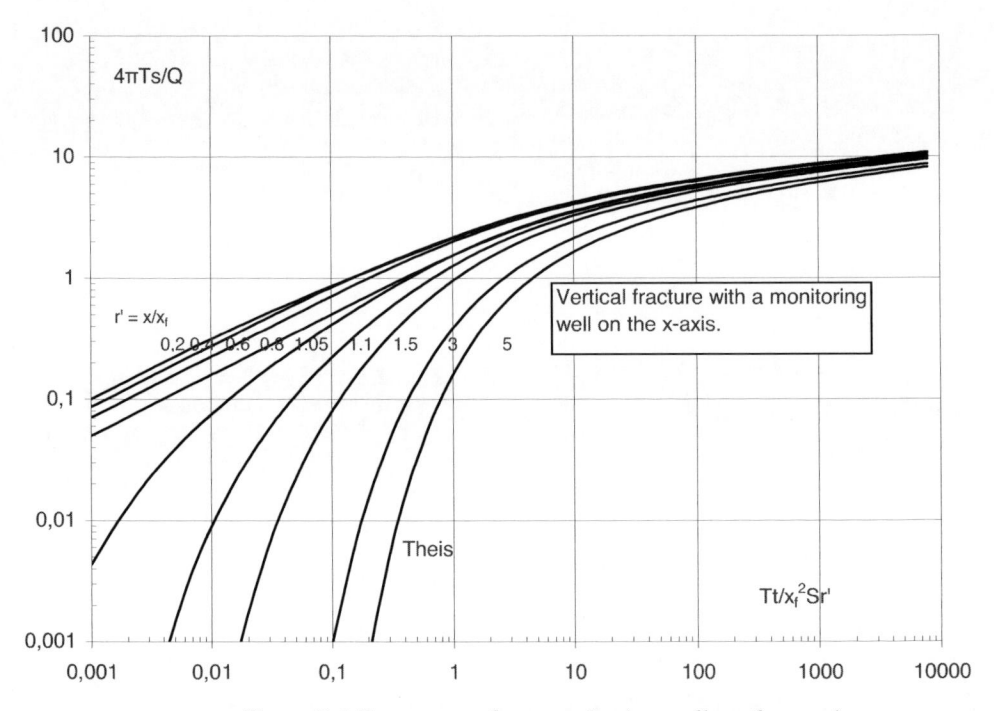

*Figure 7.4 Type curves for a monitoring well on the x-axis.*

For a monitoring well on the y-axis ($r' = y/x_f$) the equation is

$$s = \frac{Q}{4\pi T} \cdot \sqrt{\pi} \int_0^{t_D} erf\left(\frac{1}{2\sqrt{\tau}}\right) e^{-r'^2/4\tau} \frac{d\tau}{\sqrt{\tau}} \qquad (7.8)$$

Figure 7.5 shows the type-curve plot.

## VERTICAL FRACTURE

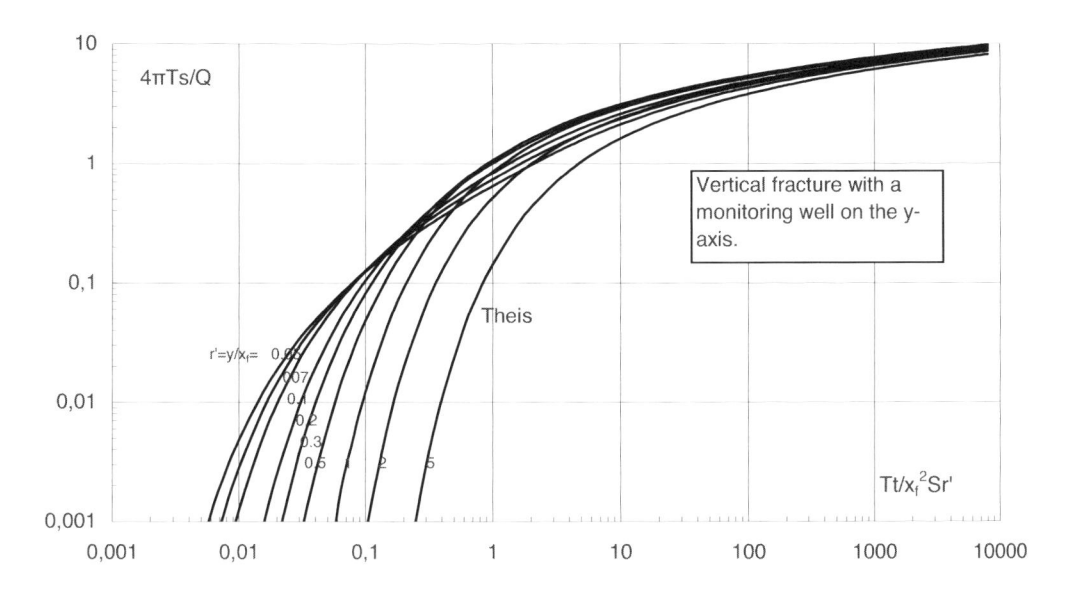

*Figure 7.5 Type curves for a monitoring well on the y-axis.*

If the monitoring well is on the 45° line ($r' = x\sqrt{2}/x_f = y\sqrt{2}/x_f$), the equation is

$$s = \frac{Q}{4\pi T} \cdot \frac{\sqrt{\pi}}{2} \int_0^{t_D} e^{-(r'/\sqrt{2})^2} \left( erf\left(\frac{1-(r'/\sqrt{2})^2}{2\sqrt{\tau}}\right) + erf\left(\frac{1+(r'/\sqrt{2})^2}{2\sqrt{\tau}}\right) \right) \frac{d\tau}{\sqrt{\tau}} \quad (7.9)$$

Figure 7.6 shows the type-curve plot.

By examining the type curves for different monitoring well locations, we realize that it may be difficult to define the correct type-curve uniquely. However, if the monitoring well is on the x-axis the drawdown looks similar to that of the pumped well if the distance is not too large. For $r' \geq 5$ all curves fall on the Theis curve. Note the displacement of the Theis curve in Figures 7.4-7.6 compared to the original Theis curve. In order to avoid errors, we use the Theis curve in the figures for calculation of parameters. For comparison, Figure 7.7 shows the original Theis curve.

# FRACTURED AQUIFERS

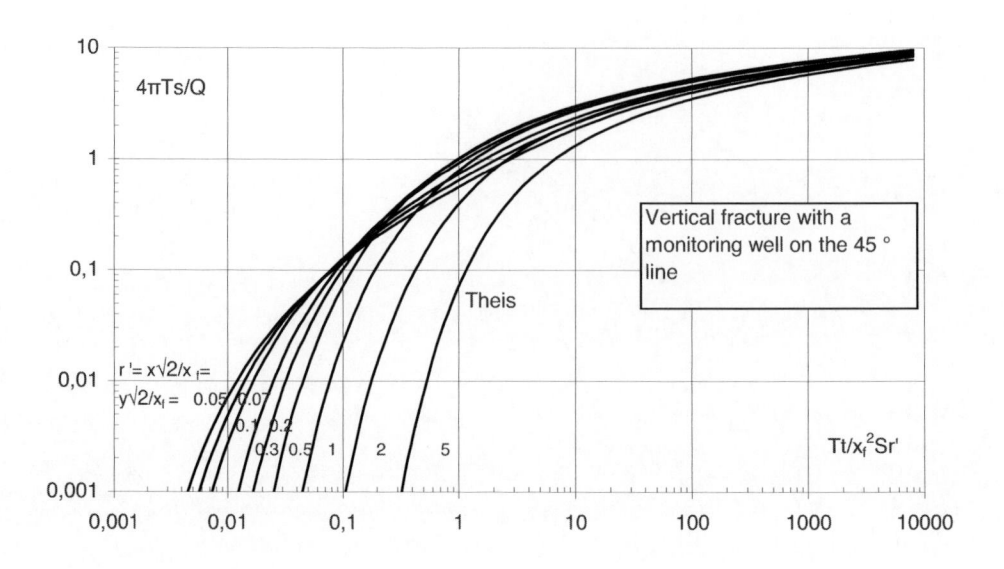

*Figure 7.6 Type curves for a monitoring well on the 45° line.*

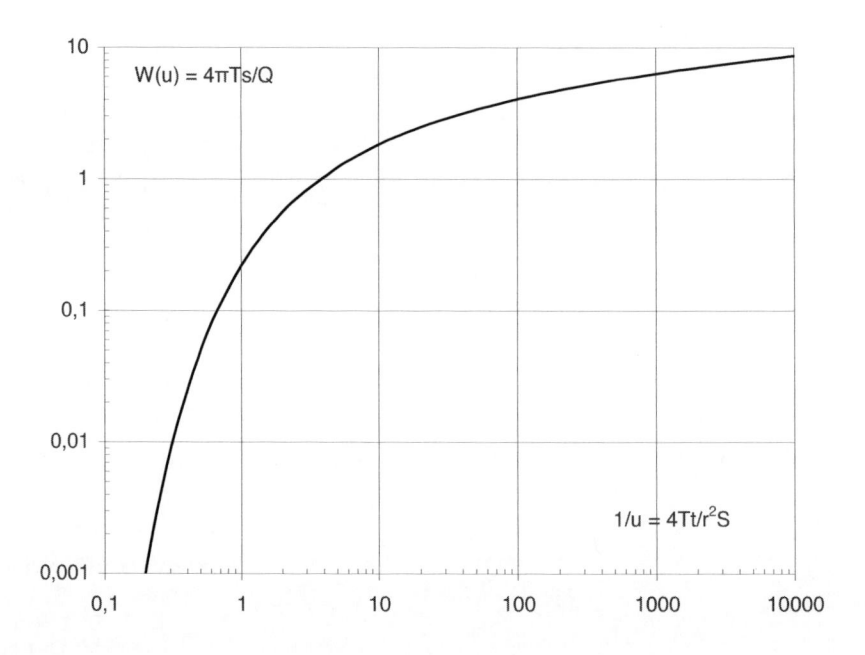

*Figure 7.7 Theis's type curve.*

## 7.2.3 Evaluation of hydraulic parameters

*The pumped well*

We use the type-curve matching procedure as usual to match data on a log-log data plot. We choose type-curve Match Points (1,1) and get the equations

$$T = \frac{Q}{4\pi s_{w\,MP}} \qquad (7.10)$$

$$x_f^2 S = Tt_{MP} \qquad (7.11)$$

From pumped well data, we can only calculate $x_f^2 S$. To find the half-fracture length $x_f$, we need to know storage from monitoring well data, or the value must be estimated.

From a semi-log data plot we get from Equation 7.6

$$T = \frac{0.183Q}{\Delta s_w} \qquad (7.12)$$

$$x_f^2 S = 996 Tt_0 \qquad (7.13)$$

where $t_0$ is obtained from the intersection with the $s_w = 0$ line. If we have well losses, the calculation of $x_f^2 S$ will be erroneous.

*Monitoring wells*

Data is plotted on log-log paper and matched with the type curves to obtain the best data fit (Figures 7.4, 7.5 or 7.6). Knowing the distance $r$ to the monitoring well we can read the value of $r'$ from the selected type-curve and determine the half-fracture length $x_f$ from $x_f = r/r'$. Using again type-curve Match Points (1,1) the following equations are obtained

$$T = \frac{Q}{4\pi s_{MP}} \qquad (7.14)$$

$$x_f^2 S = \frac{Tt_{MP}}{r'} \tag{7.15}$$

Since $x_f$ is known, storage can now be calculated.

*Gringarten and Witherspoon (1972)* presented the equations above for the anisotropic case with different transmissivity values in the x- and y-directions. If the fracture represents the x-axis then they interpreted $T_x$ as fracture transmissivity $T_f$ and $T_y$ as matrix transmissivity $T_m$.

## 7.3 The vertical fracture with infinite hydraulic conductivity

*Gringarten et al (1974)* showed that in the case of a fracture with infinite hydraulic conductivity, the drawdown in the well may de derived from the uniform flux case. When the hydraulic conductivity is infinite, the drawdown is the same throughout the fracture and equal to the drawdown in the pumped well. For $x = 0.732 x_f$ the two solutions are identical. The drawdown in the pumped well is

$$s_{wD} = \sqrt{\pi t_D}\left(erf\left(\frac{0.134}{\sqrt{t_D}}\right) + erf\left(\frac{0.866}{\sqrt{t_D}}\right)\right) - 0.067 Ei\left(-\frac{0.018}{t_D}\right) - 0.433 Ei\left(-\frac{0.750}{t_D}\right) \tag{7.16}$$

For small times, Equation 7.16 reduces to

$$s_w = \frac{Q}{2\sqrt{\pi T S x_f^2}} \sqrt{t} ,$$

i.e. the solutions for the uniform flux and infinite conductivity fracture are identical for small times. Figures 7.8 and 7.9 show Equation 7.16 together with the equation for the uniform flux fracture. It appears that the two solutions are close and it may be difficult to distinguish between them when matching data. For large times Equation 7.16 becomes *(Gringarten et al, 1974)*

$$s_{wD} = \ln t_D + 2.2 = \ln 9.03 t_D , \tag{7.17}$$

or, written with aquifer parameters

$$s_w = \frac{0.183 Q}{T} \log \frac{542 T t}{x_f^2 S} \tag{7.18}$$

Evaluation of aquifer parameters is similar to the procedure for the uniform flux fracture, but the equation for the determination of $x_f^2 S$ from a semi-log plot is different:

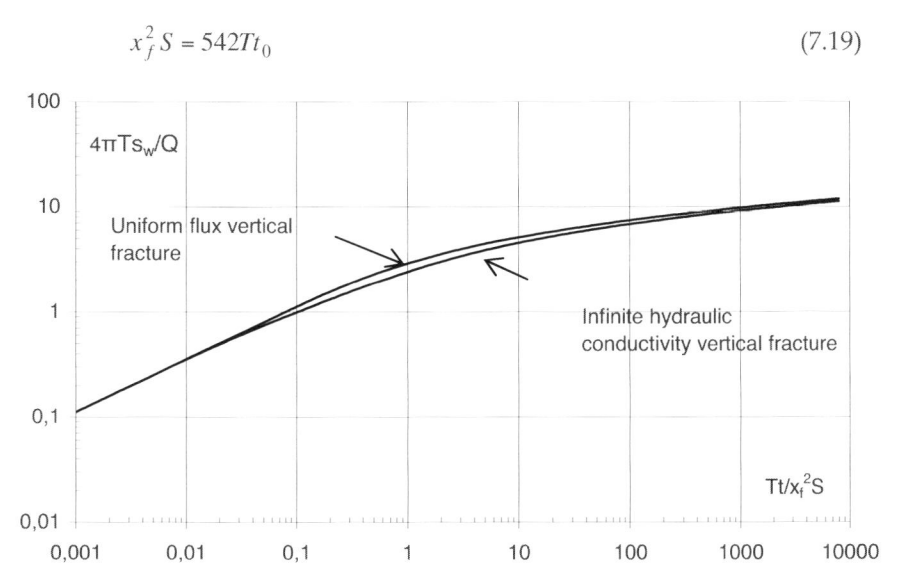

$$x_f^2 S = 542 T t_0 \qquad (7.19)$$

*Figure 7.8 Type curves for the uniform flux and infinite hydraulic conductivity fractures.*

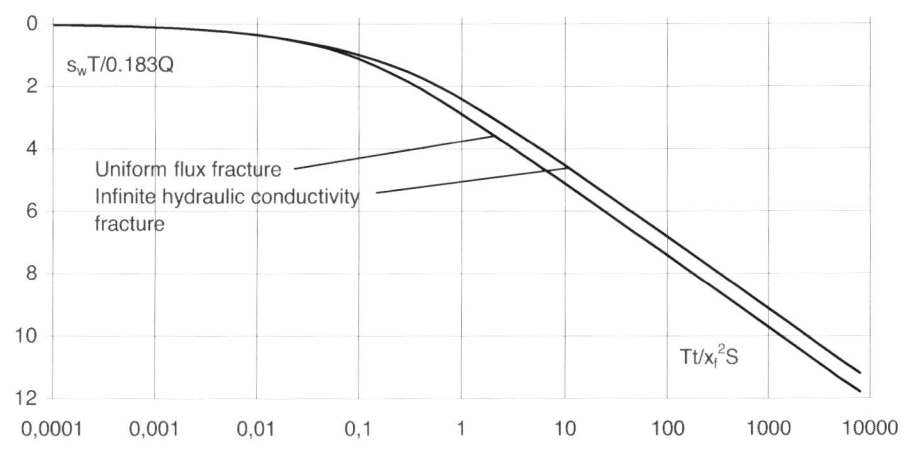

*Figure 7.9 Semi-log plot of type curves for the uniform flux and infinite hydraulic conductivity vertical fracture.*

## 7.4 Effective radius and skin factor

It is obvious that a well drilled in a vertical fracture will have en effective radius larger the well radius. *Gringarten et al (1974)* determined the effective radius for a well in a vertical fracture as shown in Figure 7.10. The aquifer is bounded by a square with half-dimension $x_e$.

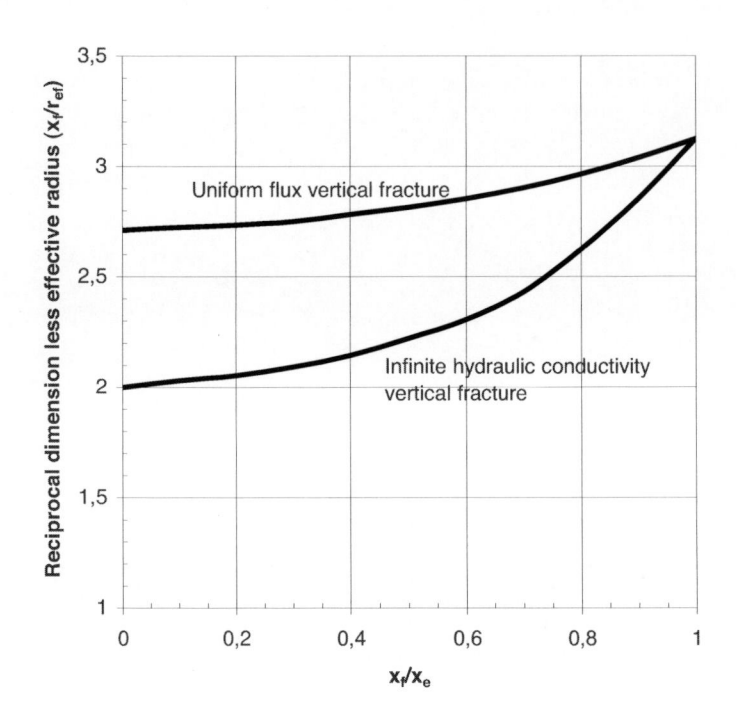

*Figure 7.10. Curves for determination of effective radius $r_{ef}$. After Gringarten et al (1974).*

Figure 7.10 shows the reciprocal dimensionless effective radius $x_f / r_{ef}$ as a function of $x_f / x_e$, where $x_f / x_e = 0$ represents an infinite aquifer. For the infinite hydraulic conductivity fracture the effective radius is exactly $½ x_f$ and for the uniform flux fracture somewhat less.

When we know the effective radius, we can determine the skin factor from Equation 4.5:

$$\sigma = -\ln \frac{r_{ef}}{r_w} \qquad (7.20)$$

# VERTICAL FRACTURE

A well in a vertical fracture will always have a negative skin factor.

Example 7.1 Analysis of data from the pumped well

The example is from *Gringarten, Ramey and Raghaven (1975)*. Data shows the recovery in an oil well. Data is $Q = 0.00097$ m³/s, $r_w = 0.076$ m, $t_p = 486,000$ min and the drawdown before stop of pumping $s_w = 241.1$ m. Figure 7.11 and the table below show recovery data.

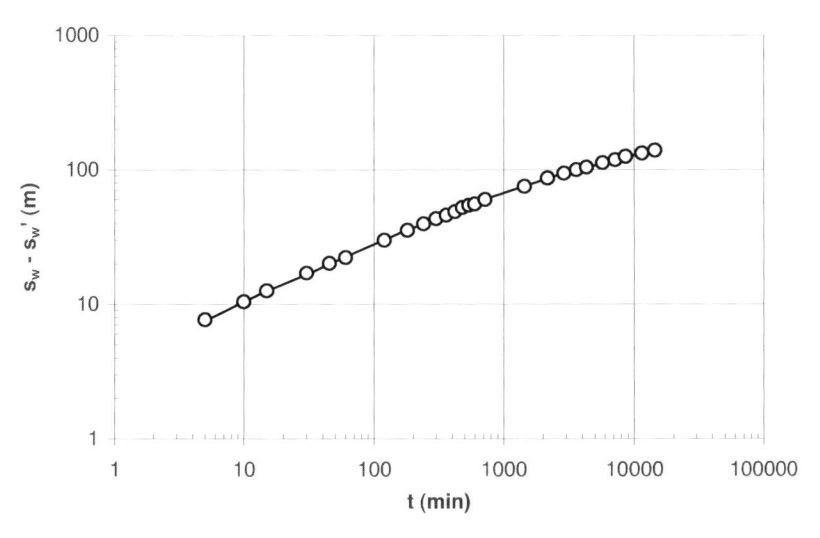

*Figure 7.11 Recovery data for Example 7.1. The curve represents the simulated drawdown.*

| Δt (min) | (Δt+t_p)/Δt | s_w-s_w' (m) | Δt (min) | (Δt+t_p)/Δt | s_w-s_w' (m) |
|---|---|---|---|---|---|
| 5 | 93600 | 7.678 | 540 | 868 | 54.444 |
| 10 | 46700 | 10.47 | 600 | 781 | 55.84 |
| 15 | 31200 | 12.564 | 720 | 651 | 60.028 |
| 30 | 15600 | 17.101 | 1440 | 326 | 75.384 |
| 45 | 10400 | 20.242 | 2160 | 218 | 86.552 |
| 60 | 7800 | 22.336 | 2880 | 164 | 94.23 |
| 120 | 3900 | 30.014 | 3600 | 131 | 99.814 |
| 180 | 2600 | 35.598 | 4320 | 109 | 104.7 |
| 240 | 1950 | 39.786 | 5760 | 82.3 | 113.076 |
| 300 | 1560 | 43.276 | 7200 | 66 | 118.66 |
| 360 | 1300 | 46.068 | 8640 | 55.2 | 125.64 |
| 420 | 1120 | 48.86 | 11520 | 41.6 | 132.62 |
| 480 | 976 | 52.35 | 14400 | 33.5 | 139.6 |

# FRACTURED AQUIFERS

Recovery data matches the type curve for a vertical fracture having infinite hydraulic conductivity. The data sheet Match Point co-ordinates are $s_{wMP} = 26$ m and $t_{MP} = 830$ min. This gives the aquifer transmissivity

$$T = \frac{Q}{4\pi s_{MP}} = \frac{0.00097}{4\pi \cdot 26} = 2.97 \cdot 10^{-6} \text{ m}^2/\text{s, and}$$

$$x_f^2 S = T t_{MP} = 2.97 \cdot 10^{-6} \cdot 830 \cdot 60 = 0.148 \text{ m}^2$$

Laboratory tests provide a storage value, $S = 0.000207$ and we can calculate the half-fracture length

$$x_f = \sqrt{\frac{x_f^2 S}{S}} = \sqrt{\frac{0.148}{0.000207}} = 26.8 \text{ m}$$

The effective radius becomes

$$r_{ef} = \frac{x_f}{2} = 13.4 \text{ m}$$

From Equation 7.20 we get the skin factor

$$\sigma = -\ln \frac{r_{ef}}{r_w} = -\ln \frac{13.4}{0.076} = -5.17$$

The skin loss is

$$s_s = \sigma \frac{Q}{2\pi T} = -5.17 \frac{0.00097}{2\pi \cdot 2.97 \cdot 10^{-6}} = -269 \text{ m}.$$

The well efficiency is

$$V = \frac{s_w - s_s}{s_w} = \frac{241.1 + 269}{241.1} = 211\%.$$

The well is stimulated because of the fracture.

We can do the same calculation based on a semi-log plot shown in Figure 7.12.

The transmissivity is determined from

$$T = \frac{0.183Q}{\Delta(s_w')} = \frac{0.183 \cdot 0.00097}{64.5} = 2.76 \cdot 10^{-6} \text{ m}^2/\text{s}$$

The skin factor is from Equation 4.18:

$$\sigma = 1.1513\left(\frac{s_w(t_p) - s_w(60)}{\Delta s_w} - \log\frac{T}{r_w^2 S} - 2.13\right) =$$

$$1.1513\left(\frac{241.1 - 260}{64.5} - \log\frac{2.76 \cdot 10^{-6} \cdot 60}{0.076^2 \cdot 0.000207} - 2.13\right) = -5.26$$

Here we have used $\Delta t = 60 \text{ min} \Rightarrow (t_p + \Delta t)/\Delta t = (486{,}000 + 60)/60 = 8101 \Rightarrow s_w' = 260$ m and $(t_p + 60)/t_p \cong 1$.

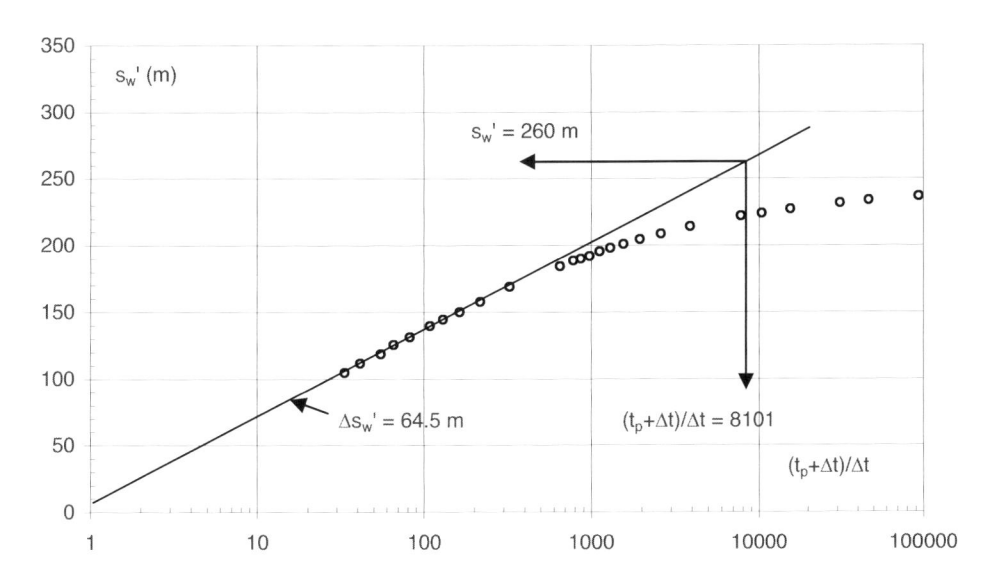

*Figure 7.12 Recovery data in a semi-log plot.*

Effective radius is

$$r_{ef} = r_w e^{-\sigma} = 0.076 \cdot e^{5.26} = 14.6 \text{ m}$$

The half-fracture length is

$$x_f = 2r_{ef} = 29.2 \text{ m}.$$

The two methods produce virtually the same results. Note that the log-log data plot enables us to identify the fracture type. This is not possible from the semi-log plot.

## 7.5 Vertical fracture with finite hydraulic conductivity

For the cases of the uniform flux and infinite conductivity vertical fractures, the fracture itself has no specific hydraulic parameters. *Cinco, Samaniego and Dominguez (1978)* published an interpretation method that allows the fracture to have a finite hydraulic conductivity and a fracture aperture. They solved the governing equations numerically; therefore, no analytical equation is available for the small time solution. The physical set-up of the fracture is similar to that of *Gringarten et al (1974)*, Figure 7.1, but now the fracture hydraulic conductivity is $K_f$ and the fracture aperture $w$. The aquifer transmissivity is $T$, the aquifer hydraulic conductivity is $K$, the aquifer thickness is $h$ and aquifer storage is $S$. The aquifer parameters represent homogeneous aquifer behavior and the aquifer itself is naturally fractured. The usual assumptions about a fully penetrating well and fracture, half-fracture length $x_f$, infinite aquifer, no well bore storage and no turbulent losses apply.

Cinco et al introduced the dimensionless parameter

$$K_{fD} w_{fD} = \frac{K_f w}{K x_f} \tag{7.21}$$

Dimensionless drawdown is as before

$$s_{wD} = \frac{4\pi T s_w}{Q}, \tag{7.22}$$

and dimensionless time is

$$t_D = \frac{Tt}{x_f^2 S} \tag{7.22}$$

Figure 7.13 shows the type-curve plot for different values of $\dfrac{K_f w}{K x_f}$.

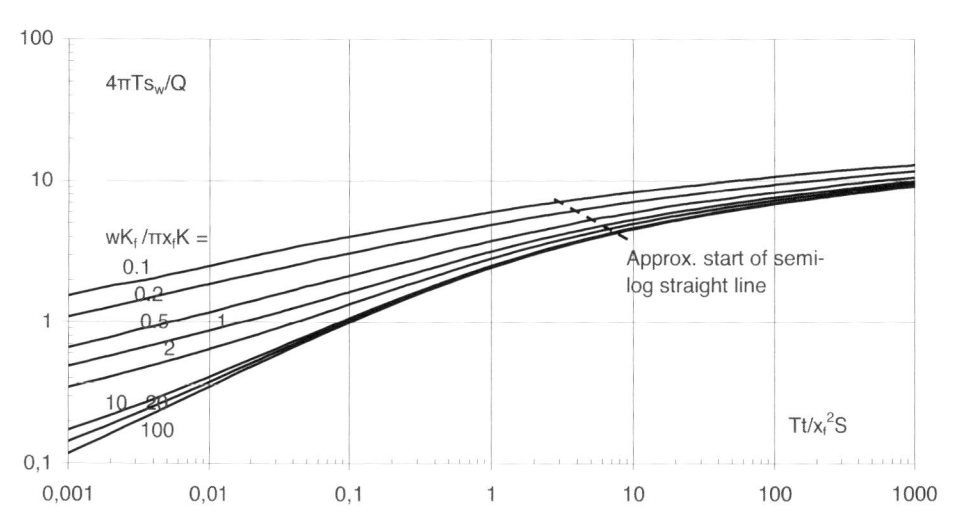

*Figure 7.13 Type curves for a vertical fracture with finite hydraulic conductivity. After Cinco et al (1978).*

The equations for evaluation of $T$ and $S$ using the Match Point procedure is

$$T = \frac{Q}{4\pi s_{wMP}} \qquad (7.23)$$

$$x_f^2 S = Tt_{MP} \qquad (7.24)$$

The fracture hydraulic conductivity $K_f$ is determined in terms of the parameter $wK_f / x_f K$ obtained from the data match. To determine the half-fracture length we again need a storage value from a monitoring well.

By comparing the type curves in Figures 7.13 and 7.2 we see that when the fracture hydraulic conductivity is large ($wK_f / x_f K \geq 100\pi$), the drawdown is equal to that for an infinite hydraulic conductivity fracture. For large values of time and the parameter $wK_f / x_f K$ we can therefore use a semi-log plot to evaluate aquifer parameters using the equations for the infinite hydraulic conductivity fracture in Chapter 7.3:

$$T = \frac{0.183Q}{\Delta s_w}, \text{ and} \qquad (7.25)$$

FRACTURED AQUIFERS

$$x_f^2 S = 542 T t_0,  \qquad (7.26)$$

where time is in minutes. Figure 7.13 indicates the validity of the straight-line method.

It appears from Figure 7.13 that for values of $wK_f / x_f K$ larger than approximately 300 the drawdown curves form a straight line with slope 1:2, but this is not the case for small and medium large values of the parameter.

*Cinco-Ley and Samaniego (1981)* presented an extended analysis of the behavior of a well intersecting a vertical finite hydraulic conductivity fracture and found that four distinct flow periods may develop: First, we see linear fracture flow, and then bilinear flow followed by formation linear flow. The final phase is pseudo-radial flow.

The bilinear period is a result of the combined linear flow in the fracture and in the aquifer. During the bilinear period, drawdown increases with $\sqrt[4]{t}$. The equation for the drawdown is

$$s_D = \frac{4.9}{\sqrt{K_{fD} w_{fD}}} \sqrt[4]{t_D} \qquad (7.27)$$

On a log-log plot of drawdown versus time, we get a line with slope ¼. If we plot drawdown against $\sqrt[4]{t}$ we also get a straight line intersecting the origin. Cinco-Ley noted that the bilinear flow period would normally to occur too early to be of practical use.

Example 7.2 Bilinear and linear flow

Data is from a recovery test on a limestone well at Roedovre west of Copenhagen, (*Nielsen, 2006 c*). Figure 7.14 shows data in a log-log plot.

We clearly see the bilinear and formation linear flow periods. The bilinear behavior also appears from Figure 7.15, which shows recovery plotted against $\sqrt[4]{\Delta t}$ for the early data, and the linear formation flow period appears from Figure 7.16 which is a plot of recovery against $\sqrt{\Delta t}$. *Cinco-Ley and Samaniego (1981)* described a method to analyze the linear and bilinear flow periods. It appears from Equation 7.27 that we need to know aquifer parameters to evaluate fracture parameters. The test in this example is too short to show aquifer parameters so the parameter evaluation is not possible. The plots in Figures 7.14 – 7.16 may serve as diagnostic tools.

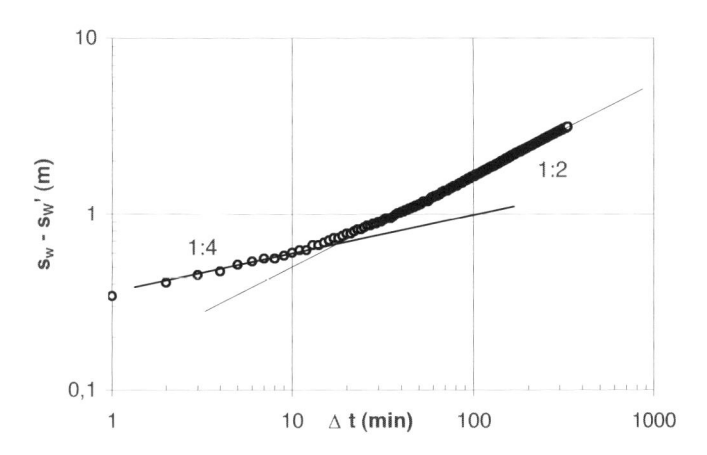

*Figure 7.14 Recovery data in a log-log plot.*

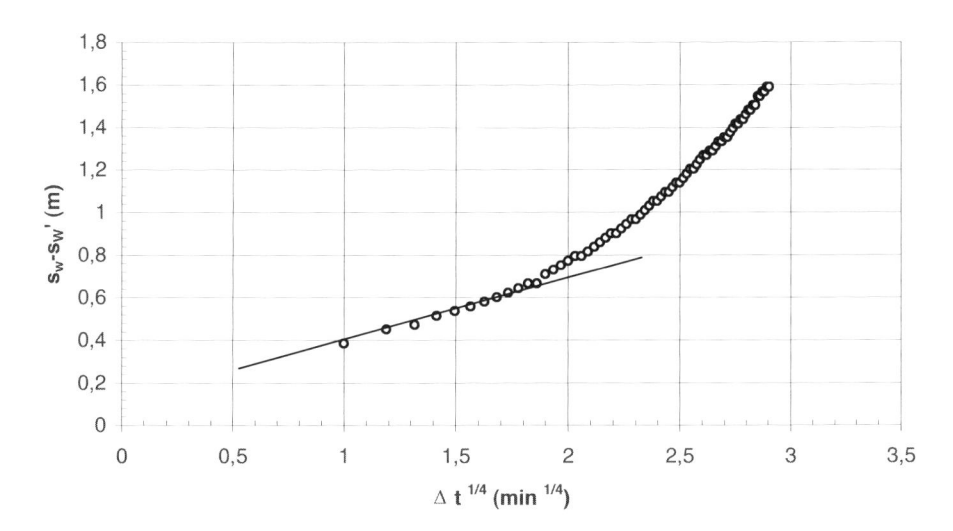

*Figure 7.15 Recovery plotted against $\sqrt[4]{\Delta t}$ .*

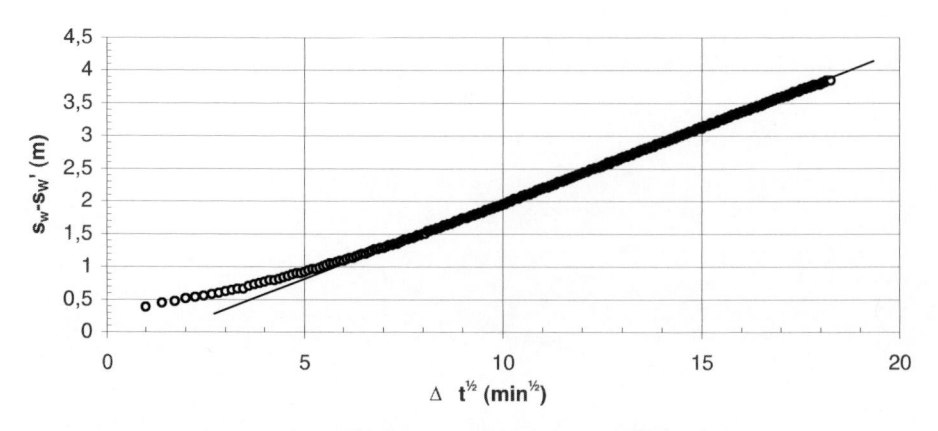

*Figure 7.16 Recovery plotted against $\sqrt{\Delta t}$.*

### 7.5.1 Effective radius and skin factor

Knowing the fracture half-length, the effective radius is determined from Figure 7.17. The effective radius depends on the parameter $K_f w / K x_f$. When this parameter is large, we approach the infinite hydraulic conductivity conditions and we get $r_{ef} = \frac{1}{2} x_f$.

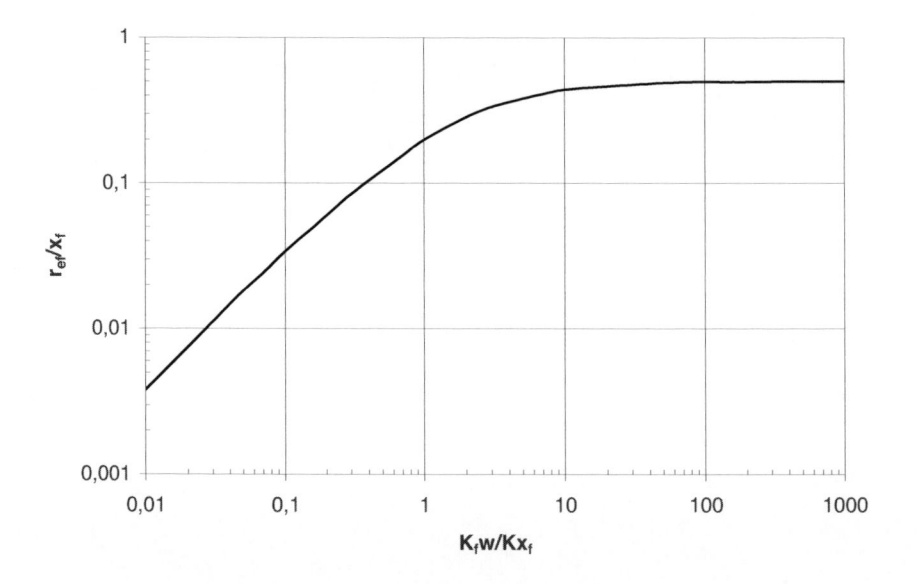

*Figure 7.17 Graph for determination of effective radius, after Cinco-Ley (1981).*

The skin factor, determined from Figure 7.18, also depends on $K_f w / K x_f$.

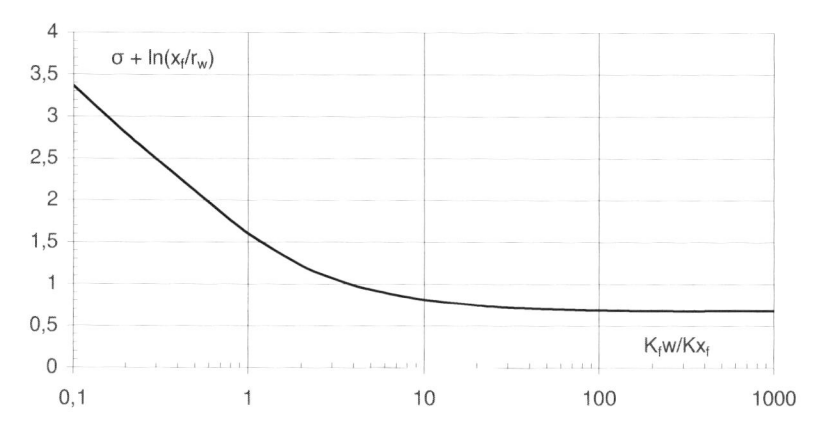

*Figure 7.18 Graph for determination of the skin factor, after Cinco et al (1978).*

7.5.2 Radius of influence for a vertical fracture

*Cinco et al (1978)* defined the radius of influence of a fracture as the distance beyond which the drawdown is similar to that of radial flow, i.e. the drawdown in monitoring wells will follow Theis curves. They found this distance to be about $4x_f$. This may be taken as a rule of thumb for all vertical fractures.

Example 7.3 Vertical fracture with finite hydraulic conductivity.

Data is from a test on a limestone well at Greve south of Copenhagen, *(Nielsen, 2004 b)*. $Q = 19.3$ m³/h, $h = 12$ m, $r_w = 0.1125$ m and $S = 0.002$ (estimate). Figure 7.19 shows the data plot and the table gives data.

The drawdown curve, Figure 7.19, matches Cinco et al's type curve. From the matching we get $s_{wMP} = 0.85$ m, $t_{MP} = 8.5$ min and $K_f w / K x_f = 0.5\pi$.

The transmissivity is from Equation 7.23,

$$T = \frac{Q}{4\pi s_{MP}} = \frac{19.3}{4\pi \cdot 0.85 \cdot 3600} = 0.0005 \text{ m}^2/\text{s}$$

and, from Equation 7.24,

$$x_f^2 S = T t_{MP} = 60 \cdot 0.0005 \cdot 8.5 = 0.255 \text{ m}^2$$

# FRACTURED AQUIFERS

| t (min) | sw (m) | t (min) | sw (m) |
|---|---|---|---|
| 0.074 | 0.923 | 60 | 4.55 |
| 0.25 | 1.32 | 90 | 4.86 |
| 0.35 | 1.49 | 120 | 5.12 |
| 0.5 | 1.63 | 180 | 5.48 |
| 0.67 | 1.718 | 240 | 5.78 |
| 1 | 1.917 | 300 | 5.97 |
| 1.67 | 2.13 | 360 | 6.15 |
| 2.5 | 2.364 | 480 | 6.405 |
| 5 | 2.698 | 600 | 6.65 |
| 6.67 | 2.84 | 720 | 6.77 |
| 10 | 3.053 | 960 | 7.01 |
| 15 | 3.32 | 1080 | 7.12 |
| 20 | 3.55 | 1200 | 7.21 |
| 30 | 3.9 | 1440 | 7.4 |

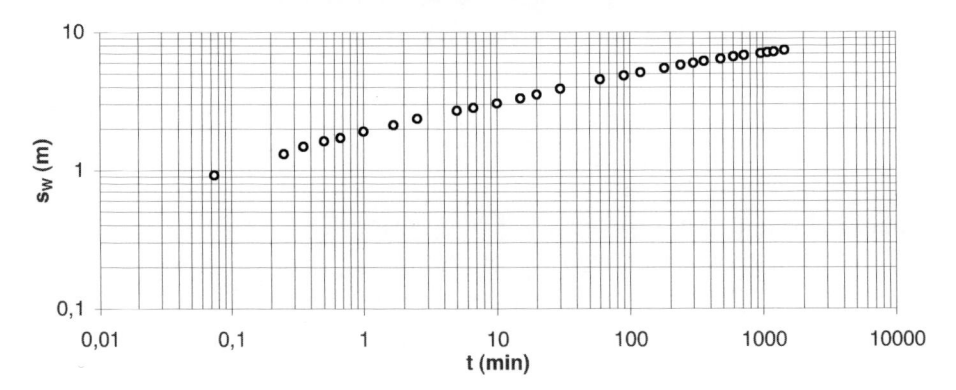

*Figure 7.19 Drawdown data from a fractured limestone well.*

The fracture half-length is:

$$x_f = \sqrt{\frac{x_f^2 S}{S}} = \sqrt{\frac{0.255}{0.002}} = 11.3 \text{ m}$$

We can calculate the hydraulic conductivity:

$$K = T/h = \frac{0.0005}{12} = 0.0000417 \text{ m/s and from this,}$$

$$K_f w = 0.5\pi K x_f = 0.5 \cdot \pi \cdot 0.0000417 \cdot 11.3 = 0.00074 \text{ m}^2/\text{s}.$$

The effective radius is determined from Figure 7.17 with $K_f w / K x_f = 0.5\pi$:

$$r_{ef} / x_f = 0.25 \Rightarrow r_{ef} = 0.25 \cdot 11.3 = 2.88 \text{ m}$$

The skin factor is determined from Figure 7.18 with $x_f / r_w = \dfrac{11.3}{0.1125} = 100.4$ and $K_f w / K x_f = 0.5\pi$:

$$\sigma + \ln(x_f / r_w) = 1.3 \Rightarrow \sigma = 1.3 - \ln 100.4 = -3.31$$

The skin loss is

$$s_s = \frac{\sigma Q}{2\pi T} = \frac{3.31 \cdot 19.3}{2\pi \cdot 0.0005 \cdot 3600} = -5.65 \text{ m}$$

and the well efficiency

$$V = \frac{s_w - s_s}{s_w} = \frac{7.40 + 5.65}{7.40} = 176\%$$

Figure 7.20 shows the simulated drawdown.

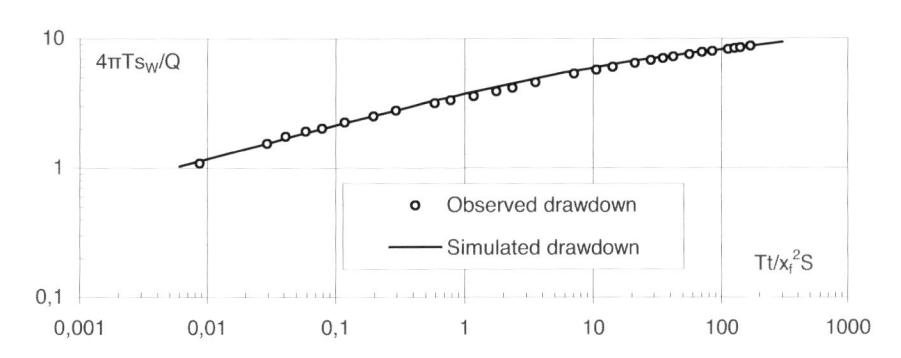

*Figure 7.20 Simulation of drawdown data plot from a limestone well.*

We can make the skin calculation the classical way from the semi-log plot shown in Figure 7.21.

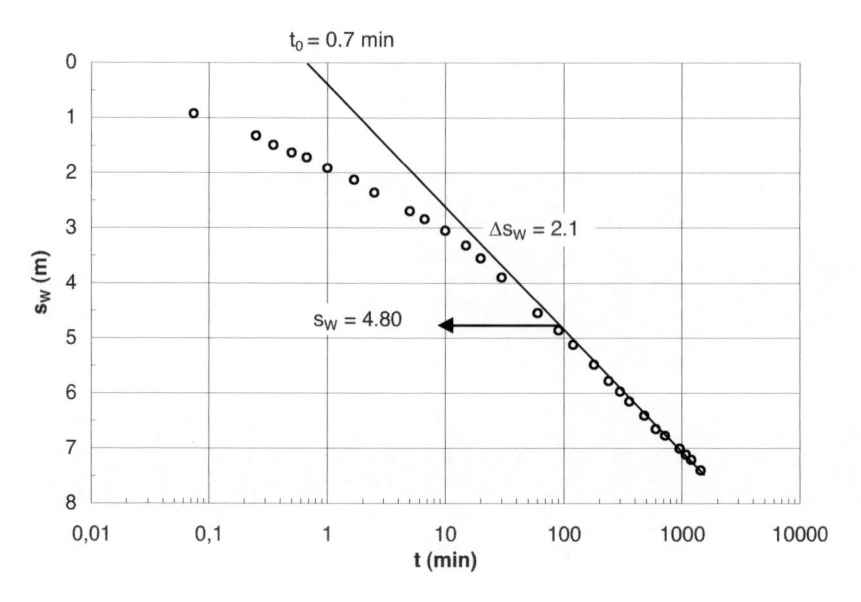

*Figure 7.21 Semi-log data plot from a limestone well.*

The transmissivity is

$$T = \frac{0.183Q}{\Delta s_w} = \frac{0.183 \cdot 19.3}{2.1 \cdot 3600} = 0.000467 \text{ m}^2/\text{s}$$

and for storage, we get from Equation 7.26:

$$x_f^2 S = 542 T t_0 = 542 \cdot 0.000467 \cdot 0.7 = 0.177 \text{ m}^2$$

and therefore the fracture half-length

$$x_f = \sqrt{\frac{x_f^2 S}{S}} = \sqrt{\frac{0.177}{0.002}} = 9.40 \text{ m}$$

The skin factor for $t = 100$ min. and $s_w = 4.80$ m becomes

$$\sigma = 1.1513\left(\frac{s_w}{\Delta s_w} - \log t - \log\frac{T}{r_w^2 S} - 2.13\right) =$$

$$1.1513\left(\frac{4.80}{2.10} - \log 100 - \log\frac{0.000467}{0.1125^2 \cdot 0.002} - 2.13\right) = -3.58.$$

The effective radius becomes

$$r_{ef} = r_w e^{-\sigma} = 0.1125 \cdot e^{3.58} = 4.04 \text{ m}$$

The skin loss is

$$s_s = \frac{\sigma Q}{2\pi T} = -\frac{3.58 \cdot 19.3}{2\pi \cdot 0.000467 \cdot 3600} = -6.54 \text{ m}$$

Finally, the well efficiency is

$$V = \frac{s_w - s_s}{s_w} = \frac{7.40 + 6.54}{7.40} = 188\%$$

The two methods are in reasonable agreement. The method based on the semi-log plot assumes a high value of $K_f w / K x_f$, which is not the case.

## 7.6 Partially penetrating vertical fractures

In the oil industry, hydraulic fracturing is a widely used method for increasing the productivity of a well. In most cases, hydraulic fracturing results in vertical fractures, which may be why vertical fractures are studied in detail. In the preceding chapters, the fracture has fully penetrated the aquifer. *Raghaven, Uraiet and Thomas (1978)* described the drawdown in a pumped well in a partially penetrating fracture. In this case, there is a vertical flow component and we need to operate with a pseudo-skin factor. The solution of Raghaven et al is very close to that of *Gringarten et al (1974)*, but since we introduce the vertical co-ordinate, equations use hydraulic conductivity and specific storage, not transmissivity and storage. Raghaven et al gave analytical solutions for the drawdown in the pumped well and presented type-curves. In principle, there are an infinite number of type-curves since the ratio between fracture height and aquifer thickness is a parameter. Methods for calculating the well productivity as a function of fracture height are given as well as a procedure to assess whether or not the fracture is only partial. Partially penetrating fractures are not included here, since no examples are available to demonstrate the method.

## 7.7 A well near a vertical fracture

*Cinco, Samaniego and Dominguez (1976)* investigated the transient drawdown behavior in a well near a fully penetrating vertical fracture with infinite hydraulic conductivity. They showed that early data display homogeneous aquifer behavior followed by a transition period with decreasing drawdown rate when the cone of depression reaches the fracture. The third and final drawdown period shows again homogeneous aquifer behavior with a transmissivity equal to that of the early data. In a semi-log plot, we get an early and a late straight line similar to the classical Warren and Root dual-porosity model. It may not be possible from pumped well data alone to decide which method of interpretation is correct. With data from monitoring wells, we have a better chance to decide whether or not we are dealing with a vertical fracture, but often the solution is not unique. This lack of uniqueness is an often-occurring problem in fractured aquifer evaluation, even when performing the well test correctly in terms of duration, monitoring frequency and selection of monitoring wells. It is therefore important - prior to model selection - to carry out diagnostics of pumped well and monitoring well data to rule out the unlikely models. Chapter 9 presents some diagnostics tools.

Example 7.4 Vertical fracture in a phreatic aquifer

The data for this example is from a well test of a fractured limestone aquifer outside the city of Koege, south of Copenhagen *(Kruger, 1998)*. Factors not included in current well-test theories influence data, and only a qualitative interpretation seems possible.

Prior to pumping, the aquifer is confined but it becomes phreatic after about 6 minutes of pumping. Then dewatering takes place. Figure 7.22 shows drawdown data. The well yield was 120 m$^3$/h.

From to 170 to 700 min, the drawdown stabilizes. This is due to recharge from a pond 600 m from the well. The pond dried completely up during the test. The drawdown curves prior to and after this period show a characteristic curvature, indicating a single fracture. The elongated shape of the cone of depression indicates that the fracture is vertical. We can check this assumption by plotting drawdown against the square root of time. This plot should produce two straight lines as shown in Figure 7.23.

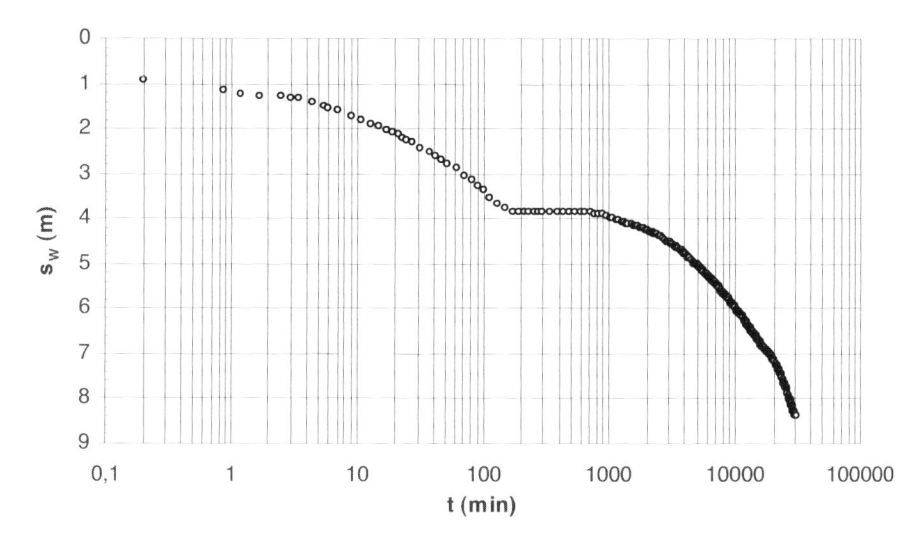

*Figure 7.22 Drawdown data from a fractured limestone well.*

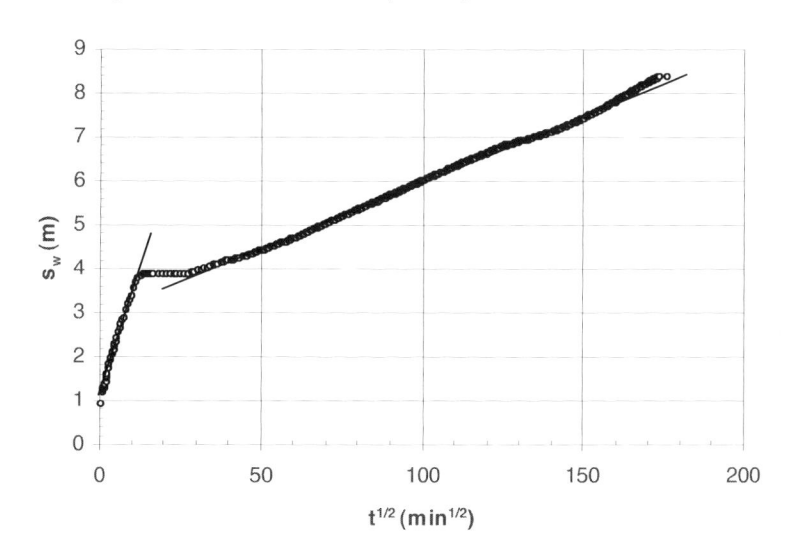

*Figure 7.23 Drawdown plotted against $\sqrt{t}$.*

Dewatering of the aquifer could be the reason for the increase in drawdown rate observed at the end of the test.

Figure 7.24 shows the residual drawdown during recovery. The recovery curve is quite different from the drawdown curve.

FRACTURED AQUIFERS

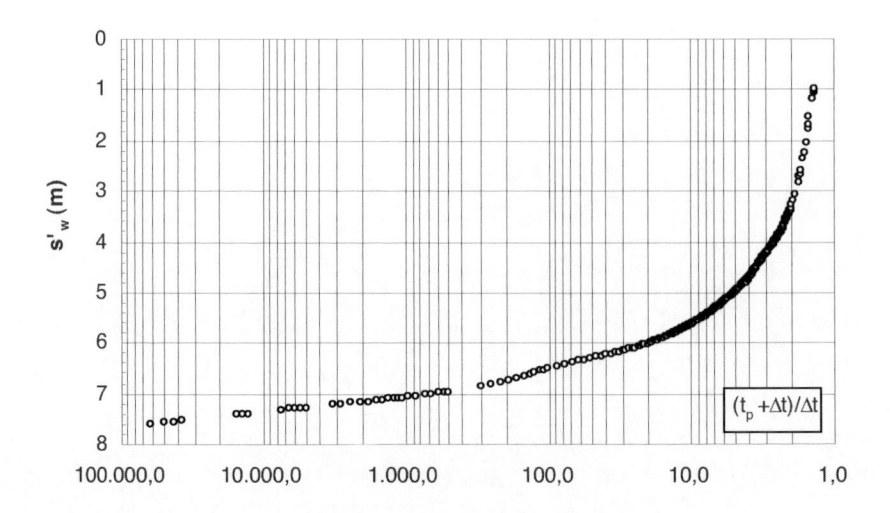

*Figure 7.24 Residual drawdown.*

We do not see the effect of the pond since the pond is not there. The main reason for the difference between drawdown and residual drawdown is that the aquifer is dewatered, and the fractures that were active during the first period of the drawdown are dry during recovery. After 80,000 min, the water level in the well has still not fully recovered.

Figure 7.25 shows residual drawdown plotted against $\sqrt{\Delta t}$. The straight line supports the single fracture assumption. The recovery curve does not show homogeneous aquifer behavior, only the linear formation flow period. Therefore, a strict data interpretation is not possible.

It is common in phreatic aquifers that drawdown data and recovery data are different, particularly when the drawdown is large. This behavior does not appear in confined aquifers, except in rare cases with pressure-dependent transmissivity.

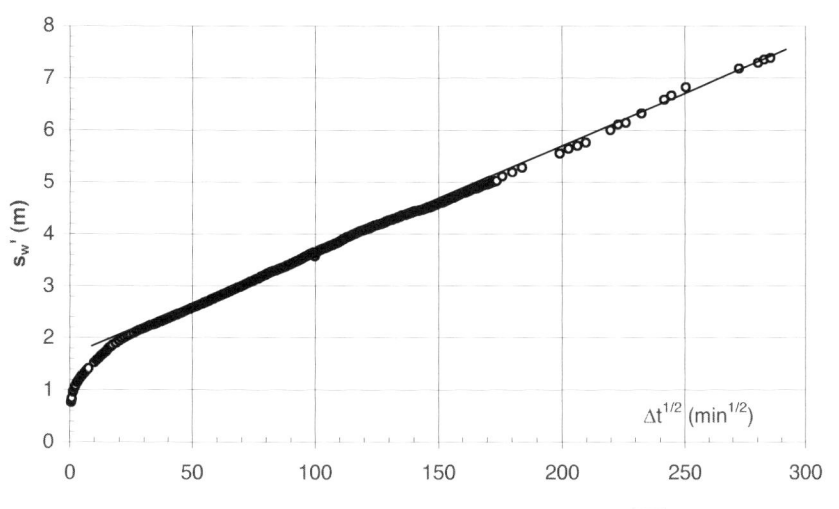

*Figure 7.25 Recovery plotted against $\sqrt{\Delta t}$.*

Example 7.5 Vertical fracture

*Rosenbeck (1974)* presented the data for this example from a well test on a fractured limestone well in central Zealand, Denmark. The well yield was 172.3 m³/hr.

| t (min) | sw (m) | t (min) | sw (m) | t (min) | sw (m) | t (min) | sw (m) |
|---|---|---|---|---|---|---|---|
| 1 | 0.771 | 9,5 | 0.886 | 120 | 1.111 | 2840 | 1.686 |
| 1.5 | 0.791 | 10 | 0.886 | 150 | 1.133 | 3260 | 1.721 |
| 2 | 0.808 | 12 | 0.902 | 180 | 1.166 | 3600 | 1.754 |
| 2.5 | 0.821 | 14 | 0.921 | 210 | 1.191 | 3960 | 1.768 |
| 3 | 0.827 | 16 | 0.931 | 240 | 1.206 | 4560 | 1.803 |
| 3.5 | 0.846 | 18 | 0.938 | 300 | 1.253 | 4600 | 1.836 |
| 4 | 0.843 | 20 | 0.946 | 360 | 1.281 | 5000 | 1.861 |
| 4.5 | 0.846 | 25 | 0.961 | 450 | 1.326 | 5600 | 1.906 |
| 5 | 0.846 | 30 | 0.976 | 600 | 1.361 | 6200 | 1.955 |
| 5.5 | 0.846 | 35 | 0.986 | 780 | 1.418 | 6680 | 1.965 |
| 6 | 0.861 | 40 | 1.001 | 940 | 1.463 | 7400 | 2.019 |
| 7 | 0.871 | 50 | 1.019 | 1180 | 1.483 | 8100 | 2.046 |
| 7.5 | 0.866 | 60 | 1.031 | 1440 | 1.531 | 9100 | 2.076 |
| 8 | 0.866 | 75 | 1.061 | 1860 | 1.581 | 9600 | 2.116 |
| 8.5 | 0.876 | 90 | 1.071 | 2100 | 1.619 | 10080 | 2.121 |
| 9 | 0.886 | 105 | 1.096 | 2500 | 1.656 | | |

The table above shows data and Figure 7.26 shows a semi-log plot of drawdown.

113

FRACTURED AQUIFERS

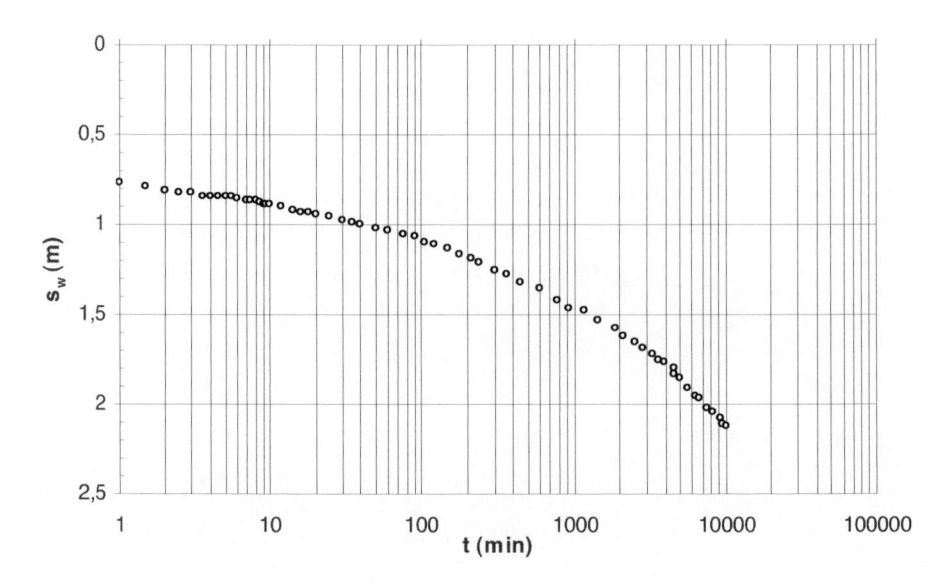

*Figure 7.26 Semi-log plot of drawdown in a fractured well.*

The drawdown curve is curvilinear and does not exhibit straight-line segments. This is an indication that the well penetrates a single fracture. Figure 7.27 shows drawdown data in a log-log plot. It has not been possible to match data with any known type-curve.

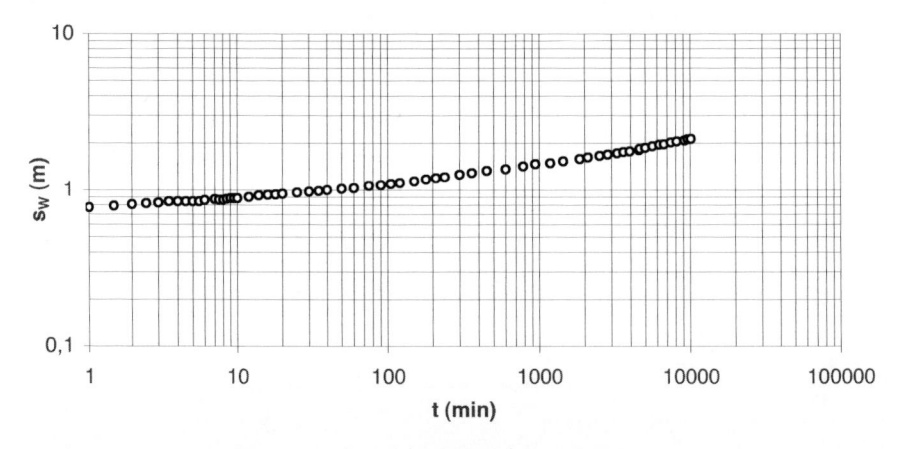

*Figure 7.27 Log-log plot of drawdown.*

Figure 7.28 shows drawdown as a function of discharge rate from a step drawdown test. Turbulent losses appear for high well yields, maybe above 0.04 m$^3$/s. Below that we see linear losses indicating a fractured well with high well efficiency.

To investigate the fracture nature of the well, we plot drawdown against $\sqrt{t}$ to identify a possible linear flow period. Figure 7.29 shows the plot. Again, data falls on a curve indicating the absence of a linear flow period.

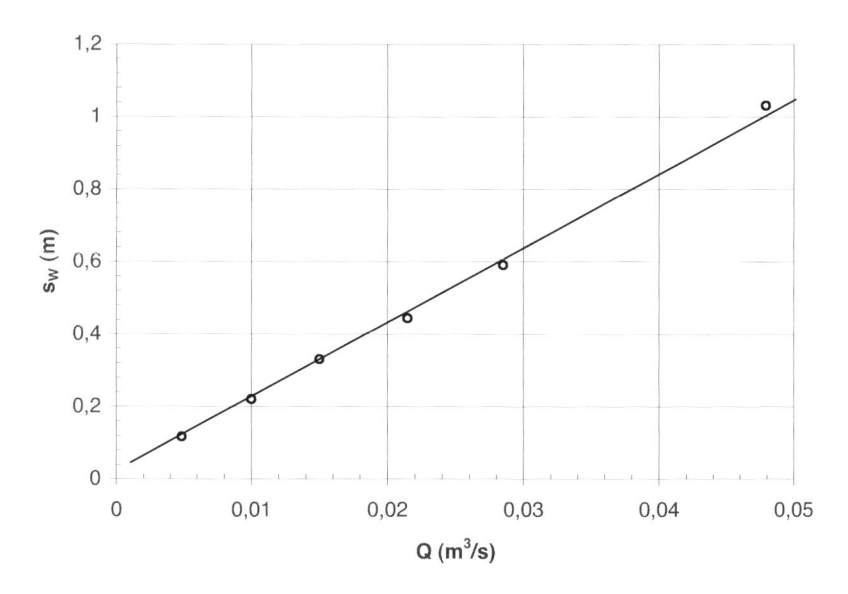

*Figure 7.28 Drawdown as a function of discharge rate.*

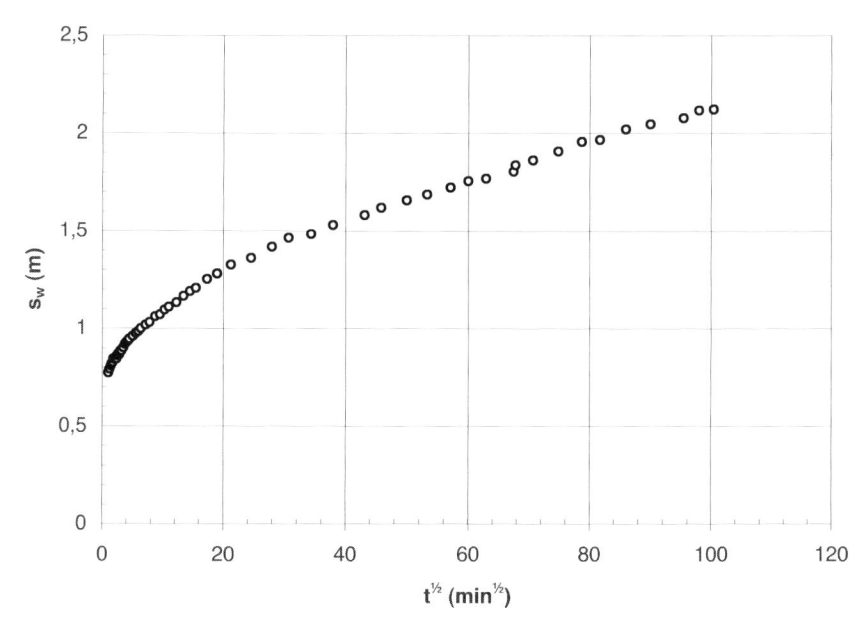

*Figure 7.29 Drawdown plotted against $\sqrt{t}$.*

Another option is to plot drawdown against $\sqrt[4]{t}$ to identify a possible bilinear flow period. Figure 7.30 shows this plot. Now data falls on a straight line showing a bilinear period during the entire test. On the log-log plot, Figure 7.27, there was no straight-line segment with slope 1:4. This is because there is a turbulent well loss that distorts data in a log-log plot, especially for small drawdowns. From Figure 7.30 we can estimate the turbulent loss at 0.63 m for $t = 0$. We can subtract 0.63 from all drawdown data and replot the adjusted drawdown. Figure 7.31 shows the log-log plot and Figure 7.32 shows drawdown plotted against $\sqrt[4]{t}$.

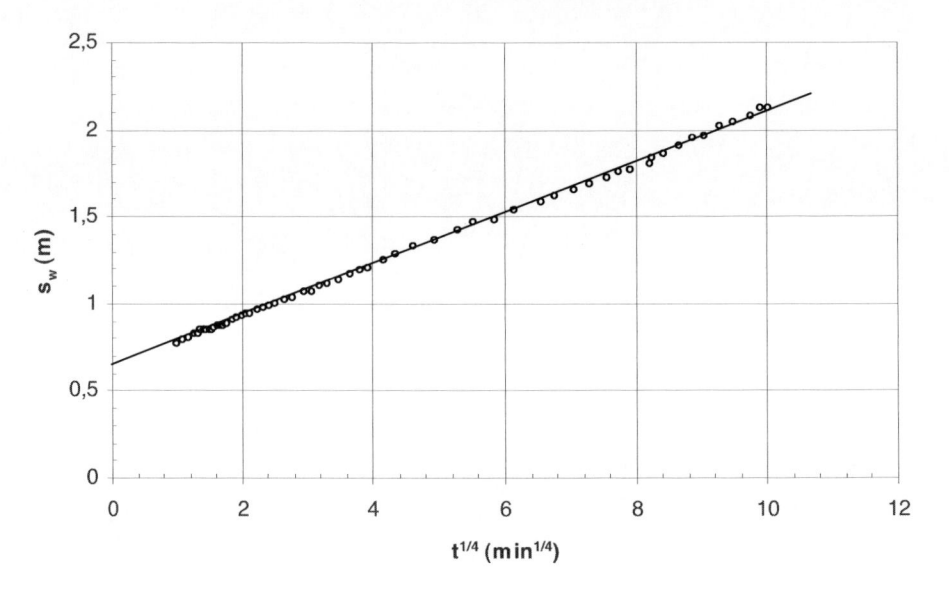

*Figure 7.30 Drawdown plotted against $\sqrt[4]{t}$.*

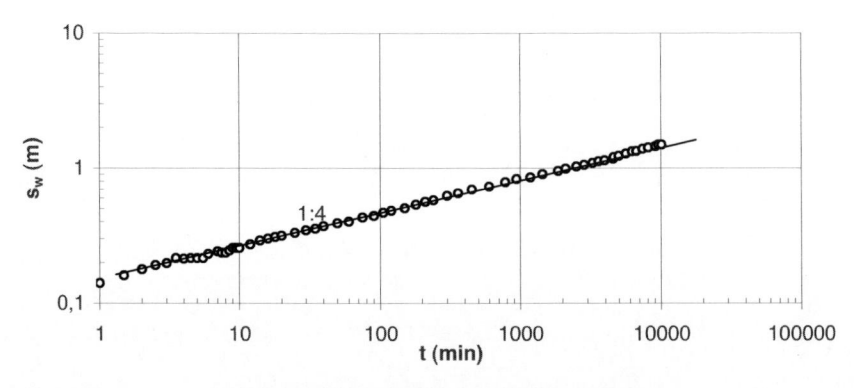

*Figure 7.31 Log-log plot of adjusted data.*

It is now evident from Figures 7.31 and 7.32 that the well exhibits bilinear flow only. It is therefore not possible to evaluate specific hydraulic parameters from drawdown data, only a combination of parameters.

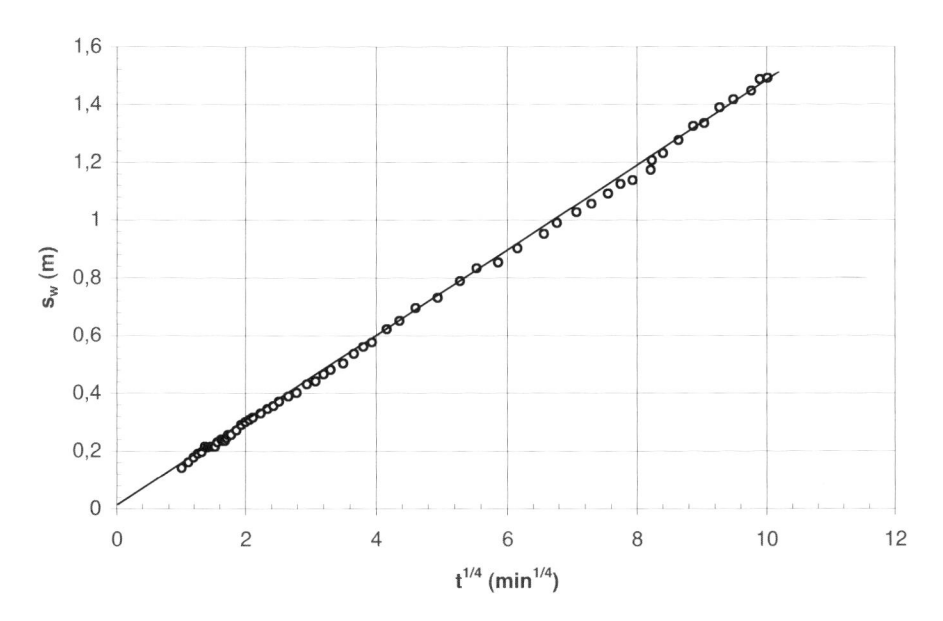

*Figure 7.32 Adjusted drawdown plotted against $\sqrt[4]{t}$*

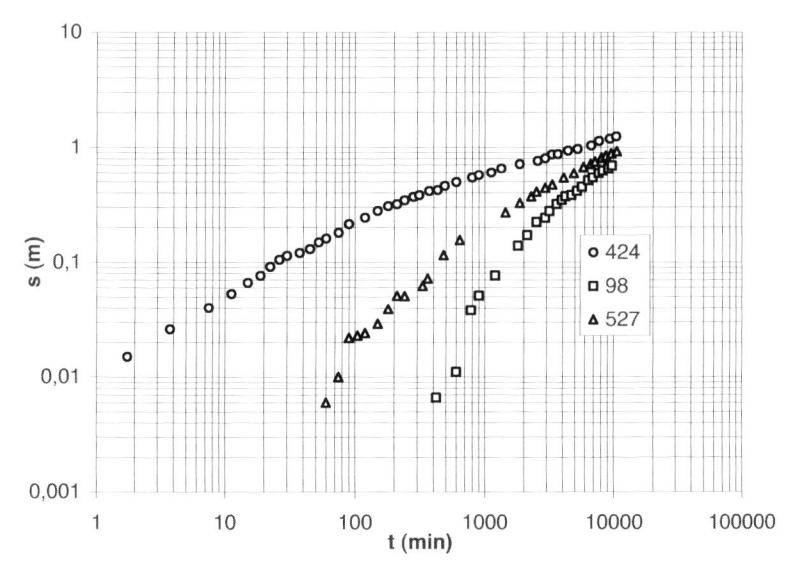

*Figure 7.33 Drawdown data from three monitoring wells.*

# FRACTURED AQUIFERS

| 424 | | | | 98 | | 527 | |
|---|---|---|---|---|---|---|---|
| t (min) | s (m) | t (min) | s (m) | t (min) | s (m) | t (min) | s (m) |
| 1.75 | 0.015 | 900 | 0.57 | 420 | 0.0066 | 60 | 0.00195 |
| 3.75 | 0.026 | 1120 | 0.6 | 600 | 0.011 | 75 | 0.01 |
| 7.5 | 0.04 | 1340 | 0.65 | 780 | 0.038 | 90 | 0.0219 |
| 11.25 | 0.053 | 1860 | 0.71 | 900 | 0.051 | 105 | 0.0228 |
| 15 | 0.066 | 2570 | 0.76 | 1200 | 0.076 | 120 | 0.024 |
| 18.75 | 0.076 | 2940 | 0.8 | 1800 | 0.138 | 150 | 0.029 |
| 22.25 | 0.091 | 3310 | 0.86 | 2130 | 0.171 | 180 | 0.039 |
| 26.25 | 0.105 | 3680 | 0.87 | 2520 | 0.222 | 210 | 0.051 |
| 30 | 0.113 | 4440 | 0.93 | 2940 | 0.241 | 240 | 0.0505 |
| 37.5 | 0.12 | 5280 | 0.96 | 3180 | 0.276 | 330 | 0.062 |
| 45 | 0.13 | 6720 | 1.03 | 3600 | 0.318 | 360 | 0.072 |
| 52.5 | 0.148 | 7740 | 113 | 3960 | 0.345 | 480 | 0.115 |
| 60 | 0.16 | 9420 | 1.18 | 4230 | 0.37 | 640 | 0.155 |
| 75 | 0.18 | 10500 | 1.23 | 4710 | 0.38 | 1440 | 0.268 |
| 90 | 0.214 | | | 5220 | 0.412 | 1860 | 0.325 |
| 120 | 0.242 | | | 5700 | 0.45 | 2280 | 0.37 |
| 150 | 0.278 | | | 6360 | 0.51 | 2520 | 0.408 |
| 180 | 0.305 | | | 6900 | 0.54 | 2940 | 0.44 |
| 210 | 0.318 | | | 7710 | 0.59 | 3330 | 0.47 |
| 240 | 0.342 | | | 8280 | 0.62 | 4110 | 0.54 |
| 280 | 0.368 | | | 9120 | 0.65 | 4920 | 0.59 |
| 310 | 0.38 | | | 9720 | 0.69 | 5850 | 0.67 |
| 370 | 0.412 | | | | | 6660 | 0.71 |
| 430 | 0.422 | | | | | 7200 | 0.75 |
| 490 | 0.458 | | | | | 8100 | 0.81 |
| 600 | 0.495 | | | | | 8760 | 0.84 |
| 800 | 0.54 | | | | | 9600 | 0.88 |
| | | | | | | 10620 | 0.92 |

Only vertical fractures produce bilinear flow. Data from monitoring wells cab therefore be analysed using the type curves shown in Chapter 7.2.2.

Figure 7.33 shows drawdown data from three monitoring wells, wells nos. 424 and 527 situated along the fracture, and well no. 98 on a nearly 45° line. The table above shows data. The distances measured along the fracture are 610 m, 1630 m and 1580 m, respectively.

# VERTICAL FRACTURE

Data matches Gringarten and Witherspoon's type curves for monitoring wells associated with a vertical fracture. Type-curve matching gives the following results shown in the table below.

| Well No. | $s_{wMP}$ (m) | $t_{MP}$ (min) | $r'$ |
|---|---|---|---|
| 424 | 0.23 | 260 | 1.05 |
| 527 | 0.49 | 5000 | 1.2 |
| 98 | 0.5 | 2300 | 2 |

Calculation of hydraulic parameters using Equations 7.14 and 7.15 gives:

Well No. 424

$$T = \frac{Q}{4\pi s_{MP}} = \frac{172.3}{4\pi \cdot 0.23 \cdot 3600} = 0.0166 \text{ m}^2/\text{s}$$

$$x_f^2 S = \frac{Tt_{MP}}{r'} = \frac{0.0166 \cdot 260 \cdot 60}{1.05} = 246.6 \text{ m}^2$$

$$x_f = \frac{x}{r'} = \frac{610}{1.05} = 581 \text{ m}$$

$$S = \frac{x_f^2 S}{x_f^2} = \frac{246.6}{581^2} = 0.00073$$

Well No. 527

$$T = \frac{Q}{4\pi s_{MP}} = \frac{172.3}{4\pi \cdot 0.49 \cdot 3600} = 0.00778 \text{ m}^2/\text{s}$$

$$x_f^2 S = \frac{Tt_{MP}}{r'} = \frac{0.00778 \cdot 5000 \cdot 60}{1.2} = 1945 \text{ m}^2$$

$$x_f = \frac{x}{r'} = \frac{1630}{1.2} = 1358 \text{ m}$$

# FRACTURED AQUIFERS

$$S = \frac{x_f^2 S}{x_f^2} = \frac{1945}{1358^2} = 0.00105$$

Well No. 98

$$T = \frac{Q}{4\pi s_{MP}} = \frac{172.3}{4\pi \cdot 0.5 \cdot 3600} = 0.00762 \, \text{m}^2/\text{s}$$

$$x_f^2 S = \frac{T t_{MP}}{r'} = \frac{0.00762 \cdot 2300 \cdot 60}{2} = 525.8 \, \text{m}^2$$

$$x_f = \frac{x\sqrt{2}}{r'} = \frac{1580\sqrt{2}}{2} = 1117 \, \text{m}$$

$$S = \frac{x_f^2 S}{x_f^2} = \frac{525.8}{1117^2} = 0.00042$$

The calculated values differ somewhat. We may expect the length of the fracture to be about 2 km. Figure 7.34 shows the simulated drawdown curves in a dimensionless plot.

## 7.8 A well in a vertical dike

*Boonstra and Boehmer (1986)* and *Boehmer and Boonstra (1987)* analysed the drawdown in a vertical dike and in the rock formation surrounding the dike. The well fully penetrates the dike. The dike may be several meters wide, and therefore, as an approximation, a plane sink replaces the well in the analysis. For small times therefore, the flow is parallel to the plane sink and the drawdown increases with the square root of time. At intermediate times, the flow is parallel in the dyke as well as in the aquifer surrounding the dike and during this period, we observe bilinear flow. For large times, the flow is pseudo-radial. In their papers, *Boonstra and Boehmer* presented data matching their theoretical drawdown.

*Cinco and Samaniego (1981)* in principle analysed the same fracture system, but in their analysis, the width of the dike (i.e. fracture) is smaller than the diameter of the well. Therefore, they did not use the approximation of replacing the well with a plane sink.

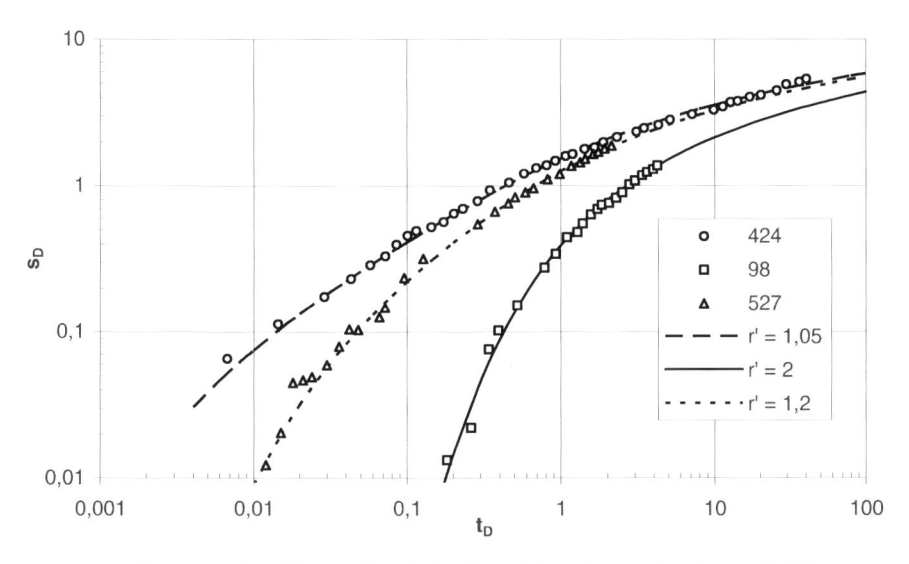

*Figure 7.34 Observed and simulated drawdowns for Example 7.5.*

## 7.9 Jenkins and Prentice

*Jenkins and Prentice (1982)* analysed the drawdown around a vertical fracture and presented the equation for drawdown in a monitoring well at distance y perpendicular to the fracture axis. The ratio of the fracture hydraulic conductivity to the aquifer hydraulic conductivity approaches infinity. The fracture length is long and its edges have no influence on the drawdown in the monitoring well. In their analysis, they did not consider bilinear flow, only linear flow. Their equation for the drawdown *s* in the monitoring well is

$$s = \frac{Q}{2LT}\left(\sqrt{\frac{4Tt}{\pi S}}e^{-u} + y\left(erf\sqrt{u} - 1\right)\right), \quad u = \frac{y^2 S}{4Tt} \tag{7.28}$$

Equation 7.28 written in dimensionless form reads

$$\frac{2\sqrt{\pi}TLs}{Qy} = \frac{e^{-u}}{\sqrt{u}} + \sqrt{\pi}\left(erf\sqrt{u} - 1\right) \tag{7.29}$$

Figure 7.35 shows the type curve, Equation 7.29.

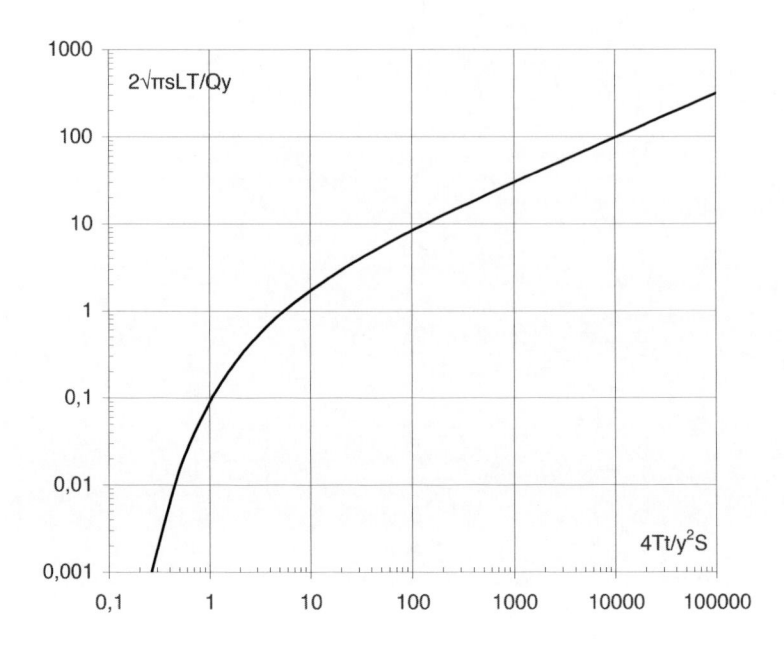

*Figure 7.35 Type curve for drawdown in a monitoring well based on Equation 7.29.*

Here, L is the fracture length. For large times, Equation 7.28 reduces to

$$s = \frac{Q}{2LT}\left(\sqrt{\frac{4T}{\pi S}}\sqrt{t} - y\right) \qquad (7.30)$$

For $y = 0$ we get the drawdown in the pumped well for large times. We see the linear flow behavior in monitoring wells as well as the pumped well.

*Edelman (1947)* analysed the flow during pumping from a well in a strip aquifer bounded by two parallel barrier boundaries. His solution is mathematically the same as Equation 7.28 but in his case, $L$ is the width of the aquifer and distance is measured parallel to the boundaries.

Example 7.6 Fracture system

Data for this example is from *Goode and Senior (1998)*. The aquifer tested is confined and the rock material is fractured mudstone and siltstone. The pumped well is MG-1125 and the monitoring well MG-1124 is 278 m from the pumped well. The tables below give drawdown data from the pumped well and the monitoring well. For the pumped well, only selected data is given.

The pumped well

| t (min) | Sw (m) | t (min) | Sw (m) | t (min) | Sw (m) |
|---|---|---|---|---|---|
| 1 | 6.77 | 25 | 914 | 300 | 11.43 |
| 2 | 6.77 | 30 | 921 | 360 | 11.55 |
| 3 | 7.01 | 35 | 9.27 | 420 | 12.1 |
| 4 | 7.16 | 40 | 9.39 | 480 | 12.31 |
| 5 | 7.25 | 50 | 9.69 | 540 | 12.44 |
| 6 | 7.32 | 60 | 10 | 600 | 12.5 |
| 7 | 7.41 | 75 | 10.18 | 720 | 12.74 |
| 8 | 7.8 | 90 | 10.36 | 850 | 12.92 |
| 9 | 8.29 | 110 | 10.55 | 1000 | 13.23 |
| 10 | 8.44 | 130 | 10.7 | 1200 | 13.78 |
| 12 | 7.87 | 150 | 10.82 | 1500 | 14.17 |
| 14 | 8.84 | 180 | 10.91 | 1800 | 15.03 |
| 16 | 8.87 | 210 | 11. | 2100 | 15.67 |
| 18 | 8.93 | 240 | 11.46 | 2400 | 16.79 |
| 20 | 9.02 | 270 | 11.4 | 2600 | 17.07 |
|  |  |  |  | 2900 | 18.47 |

Figure 7.36 shows a semi-log plot of drawdown in the pumped well and Figure 7.37 a log-log plot of drawdown in the monitoring well. This example demonstrates the ambiguity in data analysis from fractured well tests.

Monitoring well

| t (min) | s (m) | t (min) | s (m) | t (min) | s (m) |
|---|---|---|---|---|---|
| 1 | 0.0083 | 30 | 0.18 | 350 | 0.65 |
| 2 | 0.019 | 35 | 0.2 | 400 | 0.7 |
| 3 | 0.045 | 40 | 0.21 | 500 | 0.8 |
| 4 | 0.049 | 45 | 0.22 | 600 | 0.88 |
| 5 | 0.062 | 50 | 0.24 | 800 | 1.1 |
| 6 | 0.069 | 60 | 0.255 | 1000 | 1.2 |
| 7 | 0.08 | 80 | 0.3 | 1500 | 1.5 |
| 9 | 0.095 | 100 | 0.34 | 2000 | 1.7 |
| 10 | 0.099 | 150 | 0.4 | 2500 | 1.8 |
| 15 | 0.13 | 200 | 0.49 | 3000 | 1.9 |
| 20 | 0.15 | 250 | 0.58 | 4000 | 2.1 |
| 25 | 0.165 | 300 | 0.61 | 4200 | 2.3 |

# FRACTURED AQUIFERS

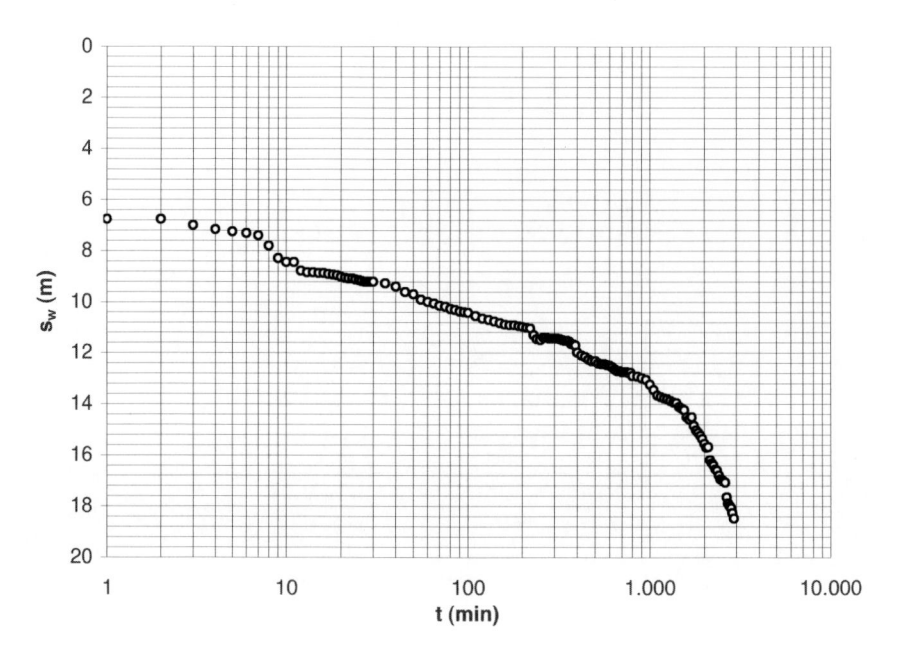

*Figure 7.36 Drawdown data from the pumped well.*

We will examine the monitoring well data plot first. After a few minutes, the drawdown increases as a function of $\sqrt{t}$. Data matches the type curve, Figure 7.35, indicating that the method of Jenkins and Prentice applies. Figure 7.38 shows the data match. Next, inspecting the data from the pumped well in a semi-log plot, Figure 7.36, we see a straight line up to about 400 min, indicating radial flow. After 400 min, the drawdown rate increases rapidly.

Figure 7.39 shows a plot of drawdown in the pumped well against $\sqrt{t}$. We see a late straight line indicating linear flow. However, Jenkins and Prentice's model only applies to a thin fracture, not a fracture system of finite width. Therefore, this interpretation is not valid.

We may consider using Edelman's model for a strip aquifer. In this case, the barrier boundaries influence drawdown data from 400 min onwards. This model of interpretation is not very likely from a geological point of view. Several other well tests of the aquifer give no indication of linear flow. On the contrary, test results indicate a dual-porosity aquifer, which is consistent with the geology. We can therefore try to match monitoring well data with Streltsova's type-curves for a dual-porosity aquifer. Figure 7.40 shows the match with late data. Matching with early

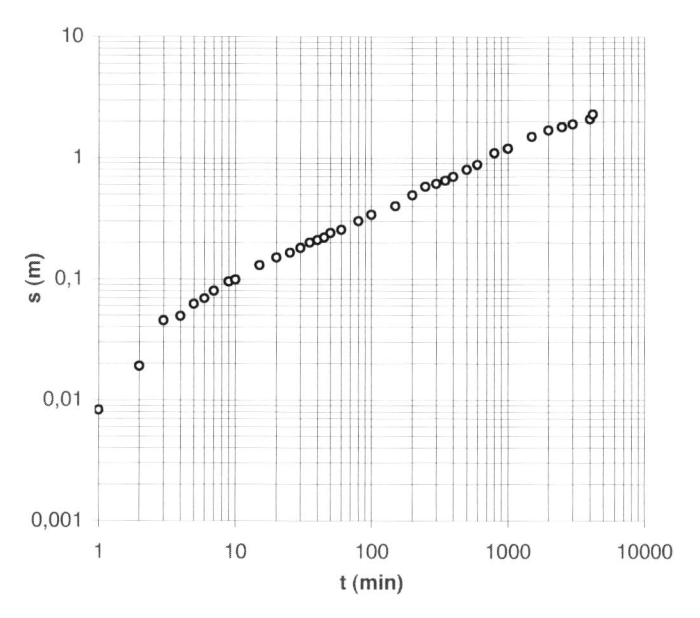

*Figure 7.37 Drawdown data from the monitoring well.*

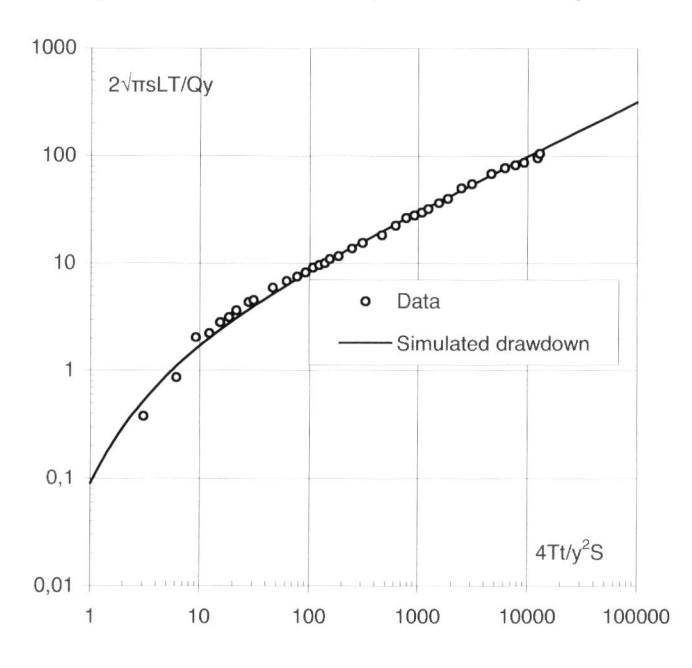

*Figure 7.38 Simulation of drawdown in the monitoring well.*

FRACTURED AQUIFERS

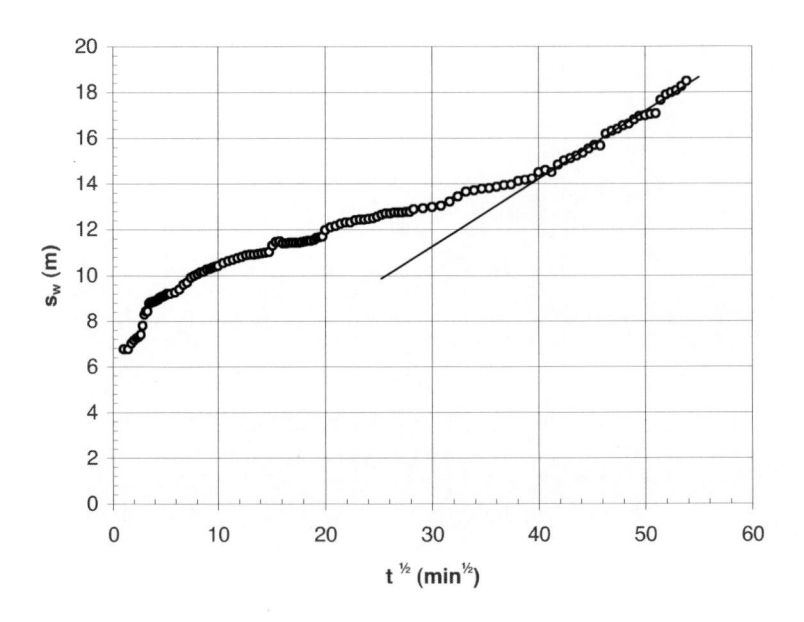

*Figure 7.39 Drawdown in the pumped well plotted against $\sqrt{t}$.*

data indicates a value of $\omega = 0.043$ showing that 4.3 % of groundwater flows in fractures. This interpretation seems more in agreement with the geology than the two previous interpretations. The consequence of this is that data from the pumped well shall also show dual-porosity behavior. Therefore, the straight-line section in Figure 7.36 is in reality the transition curve between early fracture flow and late combined fracture and matrix flow. Calculation of transmissivity of the pumped well must therefore use the log cycle slope of a line intersecting late data as shown in Figure 7.41.

The log cycle slope is $\Delta s_w = 20$ m and the well yield is 46.2 m³/hr. The transmissivity becomes

$$T_f = \frac{0.183 Q}{\Delta s_w} = \frac{0.183 \cdot 46.2}{20 \cdot 3600} = 0.000117 \text{ m}^2/\text{s}$$

From matching with monitoring well data we get $s_{MP} = 0.63$ m, $t_{MP} = 94$ min and $r/B = 2$ using type curve Match Point co-ordinates (1,1). We get the following hydraulic parameters,

126

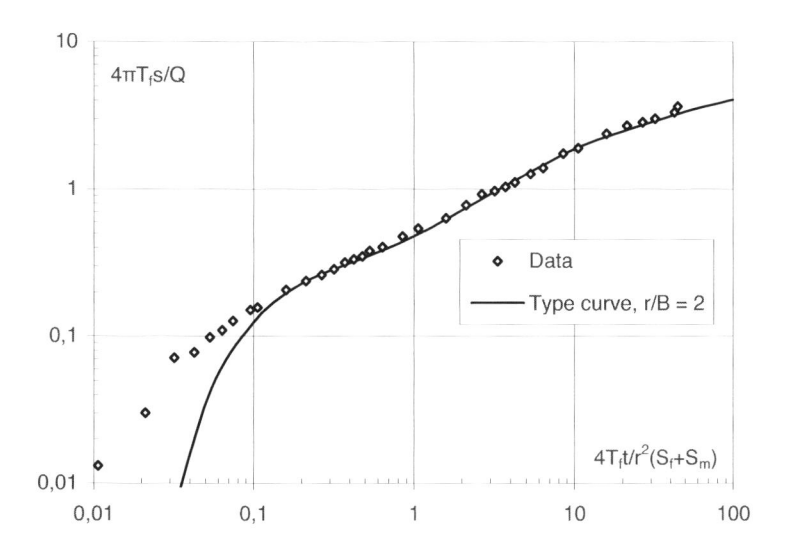

*Figure 7.40 Simulation of drawdown in the monitoring well.*

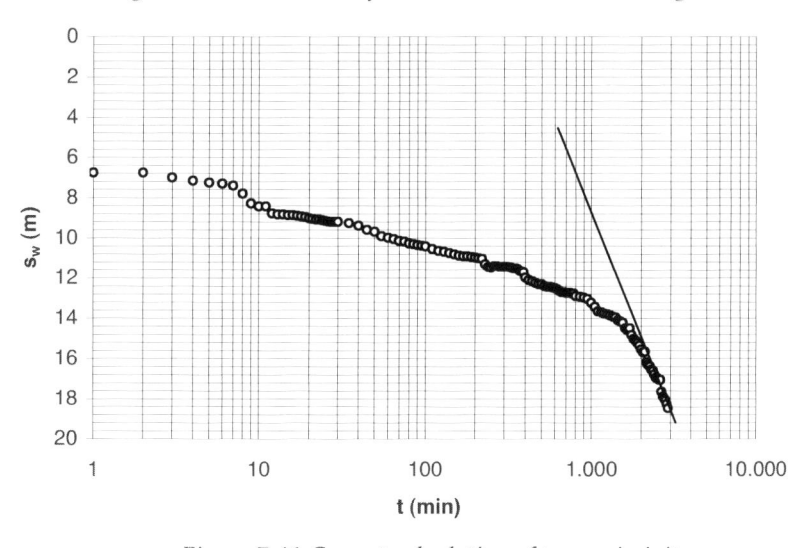

*Figure 7.41 Correct calculation of transmissivity.*

$$T_f = \frac{Q}{4\pi s_{MP}} = \frac{46.2}{4\pi \cdot 0.63 \cdot 3600} = 0.00162 \text{ m}^2/\text{s}$$

$$S_f + S_m = \frac{240 T_f t_{MP}}{r^2} = \frac{240 \cdot 0.00162 \cdot 94}{278^2} = 0.000473$$

$$S_f = \omega(S_f + S_m) = 0.043 \cdot 0.000473 = 0.0000203$$

$$\frac{K_m}{h_m} = \frac{(r/B)^2 T_f}{2r^2} = \frac{2^2 \cdot 0.00162}{2 \cdot 278^2} = 4.2 \cdot 10^{-8} \text{ s}^{-1}$$

# 8. Single horizontal fracture

Single horizontal fractures seem to be rare in oil reservoirs and the literature on such fractures is therefore limited. They seem to be common in shallow fractured carbonate aquifers and therefore evaluation methods are included here in detail. The main literature on the subject is the fundamental work of *Gringarten and Ramey (1974)*. *Gringarten and Witherspoon (1972)* and *Gringarten, Ramey and Raghaven (1975)* described the method of data analysis using practical examples.

## *8.1 The physical model*

Figure 8.1 shows the horizontal fracture set-up. The fracture is of the uniform flux type. The fracture is circular with radius $r_f$ and located at the center of formation. No specific hydraulic parameters are associated with the fracture itself. The horizontal aquifer hydraulic conductivity is $K_r$ and the vertical hydraulic conductivity is $K_z$. For practical purposes, $K_r$ is the horizontal hydraulic conductivity of the fractured aquifer system, and $K_z$ is the matrix hydraulic conductivity. The pumped well penetrates the fracture at its center, water enters the well through the fracture alone, and the fracture receives water from the surrounding aquifer. There is no well bore storage and no turbulent losses. The aquifer thickness is $h$.

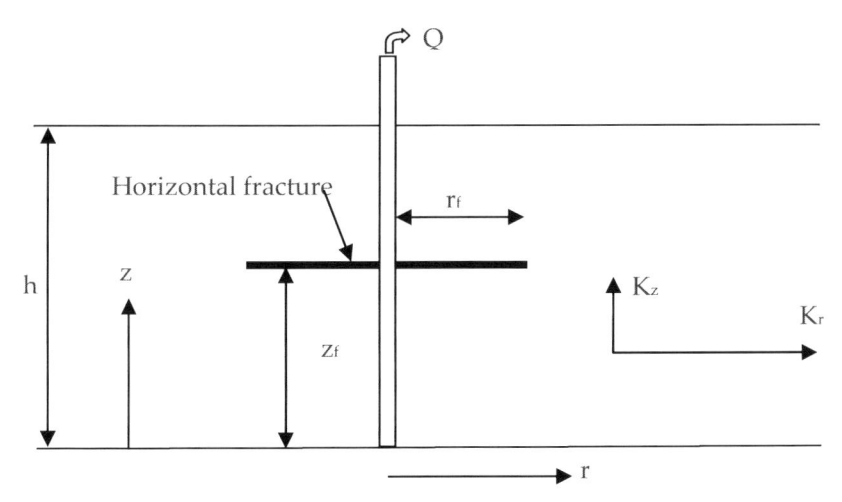

*Figure 8.1 Horizontal fracture in an anisotropic aquifer.*

The derivation of equations and solutions are complicated and not repeated here. The following dimensionless parameters are used:

Dimensionless drawdown: $\dfrac{s_{wD}}{h_D} = \dfrac{4\pi\sqrt{K_r K_z}\, r_f s_w}{Q}$ (8.1)

Dimensionless time: $t_D = \dfrac{K_r t}{S_s r_f^2}$, where (8.2)

where $S_s = S/h$ is specific storage.

Dimensionless aquifer thickness: $h_D = \dfrac{h}{r_f}\sqrt{\dfrac{K_r}{K_z}}$, (8.3)

In Equations 8.1 – 8.3 we have four unknowns, $K_r$, $K_z$, $r_f$ and $S_s$. As was the case with vertical fractures, we need to determine storage from monitoring wells or use an estimate. Since we have a vertical flow component, there is a pseudo-skin factor describing the well loss because of this.

The drawdown curve may develop three distinct periods, see Figure 8.2.

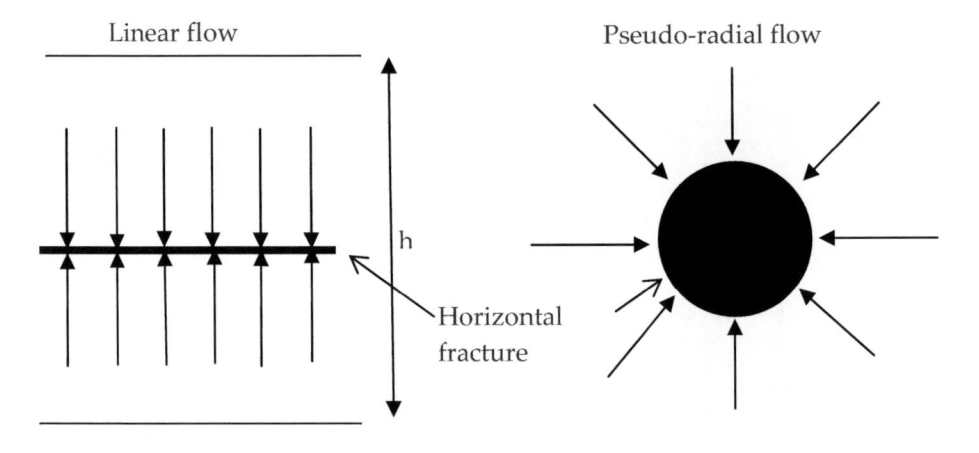

*Figure 8.2 Flow patterns for a horizontal fracture.*

For very small times, we may see radial flow in the fracture followed by linear flow in the formation. Bilinear flow cannot take place and this helps to distinguish between vertical and horizontal fractures. During the linear flow period, the flow to the fracture is vertical and the drawdown rate is proportional to $\sqrt{t}$, Figure 8.2 left. For large times, Figure 8.2 right, the flow becomes pseudo-radial and the drawdown rate becomes logarithmic. Between the linear period and the late time period there is a transition period.

Figures 8.3 and 8.4 show type-curves for the drawdown in a well in a horizontal fracture at the center of the formation. Note that dimensionless drawdown is

$$s_{wD}/h_D = \frac{4\pi\sqrt{K_r K_z}\, r_f s_w}{Q}$$

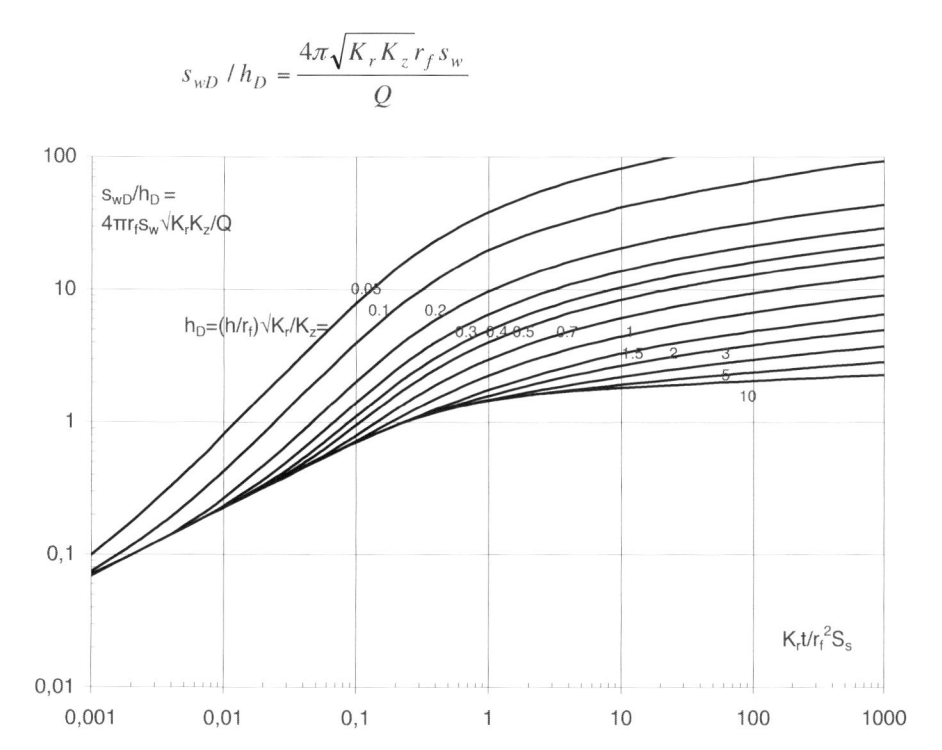

Figure 8.3 Log-log plot of type curves for the drawdown in a well in a horizontal fracture at the center of the formation. After Gringarten and Ramey (1974).

We see that for small times, the drawdown rate is proportional to $\sqrt{t}$ ( slope = ½ in Figure 8.3).

The equation for the drawdown in the pumped well for small times is (Gringarten and Ramey, 1974),

$$\frac{s_w}{h_D} = 4\sqrt{\frac{t_D}{\pi}}$$, or written in aquifer parameters

# FRACTURED AQUIFERS

$$s_w = \frac{Q}{\pi\sqrt{\pi}r_f^2\sqrt{K_zS_s}}\sqrt{t}$$

which is similar to the linear flow period for vertical fractures.

From Figure 8.3 we see that for small values of $h_D$ and for small times, the drawdown curves exhibit unit slope as is the case for a period with well bore storage. Since well bore storage is not included in the model this unit slope behavior is a result of the fracture and may result in erroneous interpretations.

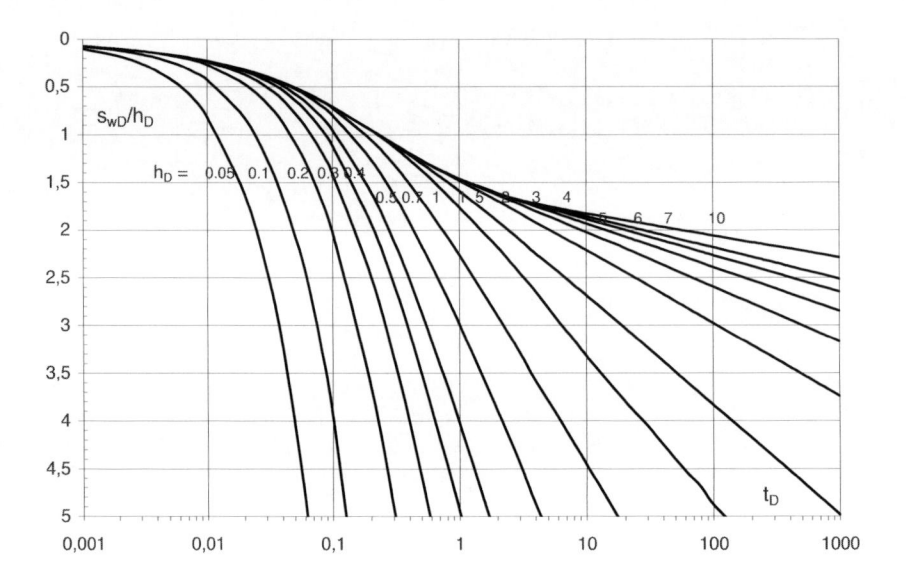

*Figure 8.4 Semi-log plot of the type curves in Figure 8.3.*

By comparing the type curve for a vertical fracture, Figure 7.2, with the type curves for a horizontal fracture, Figure 8.3, it appears that for $1 < h_D < 3$ it may be difficult to distinguish between the fracture types. For $h_D > 3$, Figure 8.4, the curves are characterized by a decreasing drawdown rate, which makes it possible to identify the fracture type. The value of $h_D$ in combination with the value of $t_D$ determines when the drawdown behaves logarithmically indicating pseudo-radial flow.

A monitoring well penetrating the fracture will have a drawdown curve similar to that of the pumped well. The drawdown in a monitoring well drilled at some distance from the fracture will follow a Theis curve. Gringarten and Ramey found that the distance $r_i$ beyond which the influence of the fracture vanishes is determined by

$$r_i = r_f + \frac{10h}{\pi}\sqrt{\frac{K_r}{K_z}} \qquad (8.4)$$

## 8.2 Evaluation of hydraulic parameters

### 8.2.1 The type-curve method for the pumped well

The hydraulic parameters $K_r$ and $K_z$ and the fracture radius $r_f$ may be evaluated using the type-curve procedure. Taking type-curve Match Point co-ordinates as (1,1) we get three equations.

From the drawdown match we get

$$\sqrt{K_r K_z}\, r_f = \frac{Q}{4\pi s_{MP}} \qquad (8.5)$$

From the time match we get

$$\frac{K_r}{S_s r_f^2} = \frac{1}{t_{MP}} \qquad (8.6)$$

From the type-curve selected, we get the third equation

$$h_D = \frac{h}{r_f}\sqrt{\frac{K_r}{K_z}} \qquad (8.7)$$

Example 8.1 demonstrates a convenient method of solving these equations. Since we have four unknowns $(K_r, K_z, S_s, r_f)$ and only three equations, one of the parameters must be determined independently. Usually $S_s$ is determined from monitoring well data.

Figure 8.3 indicates that if $h_D$ is large, i.e. the aquifer is thick, it may require a long pumping time to identify the value of $h_D$, and this may be weeks. If $h_D$ cannot be determined, the type-curve method can only provide a value of $\sqrt{K_r K_z}$.

## 8.2.2 The straight-line method for the pumped well

The straight-line method allows analysis of late data when the straight line has developed. When this is the case *Gringarten and Ramey (1974)* showed that the equation for the drawdown is

$$s_w = \frac{0.183Q}{K_r h} \log \frac{366 K_r t}{r_f^2 S_s} \tag{8.8}$$

Here we have to add the pseudo-skin loss because of the vertical flow component. Because of this loss only an approximate value of $r_f^2 S_s$ can be determined, but the straight-line method gives an accurate value of $K_r h = T$.

## 8.2.3 The straight-line method for monitoring wells

For large times monitoring well data may be used when the straight line has developed. For this case, *Gringarten and Ramey (1974)* gave the following equation for the drawdown in a monitoring well at distance $r$,

For $0 < r/r_f < 1$ the equation reads

$$s = \frac{0.183Q}{K_r h} \log \frac{366 K_r t}{r^2 S_s} \tag{8.9}$$

Here we also have a pseudo-skin loss and only $T = K_r h$ can be accurately determined.

For $r/r_f > 1$ the equation reads

$$s = \frac{0.183Q}{K_r h} \log \frac{135 K_r t}{r^2 S_s}, \tag{8.10}$$

where time is in minutes. We recognize the Jacob-Cooper equation.

### *8.3 Pseudo-skin factor, skin factor and effective radius*

The pseudo-skin factor $\sigma_p$ is proportional to the aquifer thickness and the discharge rate. $\sigma_p$ is determined from Figure 8.5. The pseudo-skin factor is always positive. The pseudo-skin loss is determined from

$$s_s = \sigma_p \frac{Q}{2\pi K_r h}$$

Skin factor, skin loss and effective radius are determined as described in Chapter 4.

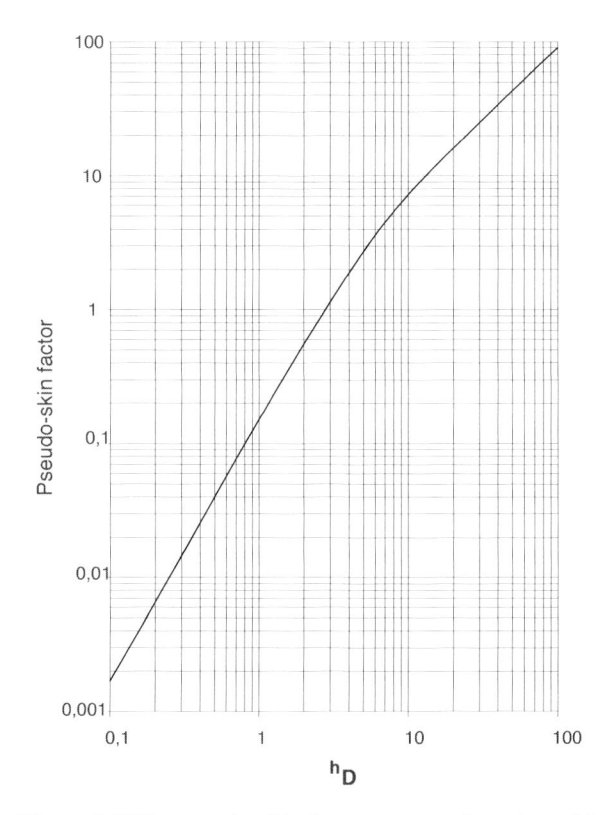

*Figure 8.5 The pseudo-skin factor $\sigma_p$ as a function of $h_D$*
*for a horizontal fracture at the center of the formation.*
*After Gringarten and Ramey (1974).*

Example 8.1

Data is from a limestone well at Gjeddesdal south of Copenhagen, *(Kruger, 2000)*. $Q = 95$ m³/hr, storage is $S = S_s h = 0.00203$ from a monitoring well, $h = 10$ m, and $r_w = 0.1125$ m. Figure 8.6 shows a plot of drawdown data and the table shows data.

FRACTURED AQUIFERS

Data matches the type curve for a single horizontal fracture with $h_D = 3$. The early linear flow period (half slope) is apparent. Evaluation of hydraulic parameters is as follows.

| t (min) | sw (m) | t (min) | sw (m) |
|---|---|---|---|
| 0.11 | 0.05 | 60 | 0.94 |
| 0.2 | 0.06 | 100 | 1.2 |
| 0.35 | 0.08 | 200 | 1.4 |
| 0.65 | 0.11 | 300 | 1.5 |
| 0.8 | 0.12 | 500 | 1.7 |
| 1 | 0.13 | 800 | 1.9 |
| 2 | 0.2 | 1000 | 2 |
| 3 | 0.22 | 2000 | 2.25 |
| 5 | 0.31 | 5000 | 2.5 |
| 10 | 0.41 | 7000 | 2.55 |
| 20 | 0..8 | 10000 | 2.65 |
| 30 | 0.68 |  |  |

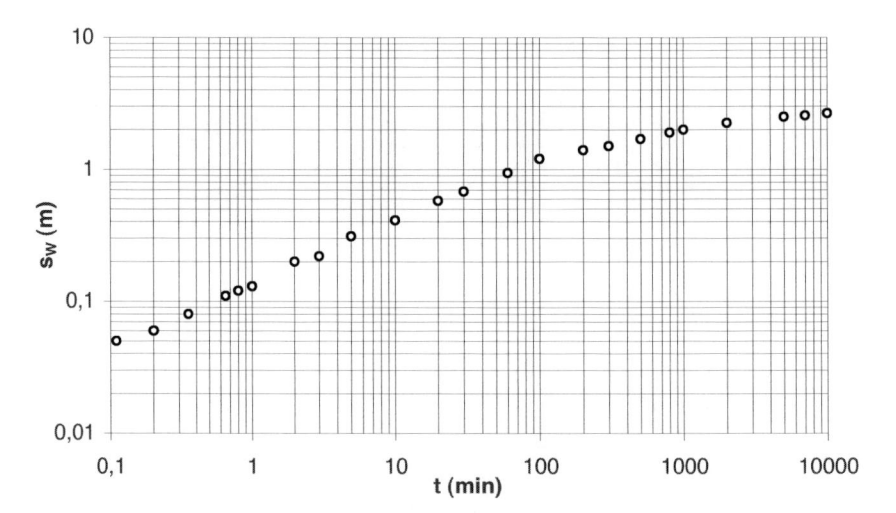

*Figure 8.6 Drawdown data from a fractured well.*

Type-curve method

The procedure follows that suggested by *Aguilera (1980)*.

1. From the data match we get the Match Point co-ordinates and the value of $h_D$:

$t_{MP} = 260$ min., $s_{wMP} = 1.0$ m and $h_D = 3$.

2. From Equation 8.5 we get

$$\sqrt{K_r K_z}\, r_f = \frac{Q}{4\pi s_{MP}} = \frac{95}{4\pi \cdot 1.0 \cdot 3600} = 0.0021 \text{ m}^2/\text{s}$$

3. From Equation 8.6 we get

$$\frac{K_r}{r_f^2} = \frac{S_s}{t_{MP}} = \frac{0.00203}{10 \cdot 260 \cdot 60} = 1.3 \cdot 10^{-8} \text{ m}^{-1}\text{s}^{-1}$$

4. From Equation 8.7 we get

$$\frac{1}{r_f}\sqrt{\frac{K_r}{K_z}} = \frac{h_D}{h} = \frac{3}{10} = 0.3 \text{ m}^{-1}$$

5. $K_r$ is determined from the identity

$$\sqrt{K_r K_z}\, r_f \cdot \frac{1}{r_f}\sqrt{\frac{K_r}{K_z}} = K_r = 0.0021 \cdot 0.3 = 0.00063 \text{ m/s}$$
$$\Rightarrow T_f = K_r h = 0.00063 \cdot 10 = 0.0063 \text{ m}^2/\text{s}$$

6. The fracture radius is determined from the relation

$$\frac{\sqrt{K_r}}{\sqrt{K_r}/r_f} = r_f = \sqrt{\frac{0.00063}{1.3 \cdot 10^{-8}}} = 220 \text{ m}$$

7. $K_z$ is determined from the relation

$$\sqrt{K_z} = \frac{\sqrt{K_r K_z}\, r_f}{\sqrt{K_r}\, r_f} = \frac{0.0021}{220 \cdot \sqrt{0.00063}} = 3.8 \cdot 10^{-4} \text{ (m/s)}^{1/2}$$

$$\Rightarrow K_z = 1.45 \cdot 10^{-7} \text{ m/s}$$

The radius of influence $r_i$ for the fracture is from Equation 8.4:

$$r_i = r_f + \frac{10h}{\pi}\sqrt{\frac{K_r}{K_z}} = 220 + \frac{10 \cdot 10}{\pi}\sqrt{\frac{0.00063}{1.45 \cdot 10^{-7}}} = 2319 \text{ m}$$

With $h_D = 3$ the pseudo-skin factor is determined from Figure 8.5 as $\sigma_p = 1.2$ and the pseudo-skin loss is

$$s_{ws} = \sigma_p \frac{Q}{2\pi K_r h} = \frac{1.2 \cdot 95}{2\pi \cdot 0.00063 \cdot 10 \cdot 3600} = 0.8 \text{ m}.$$

Figure 8.7 shows the simulated drawdown.

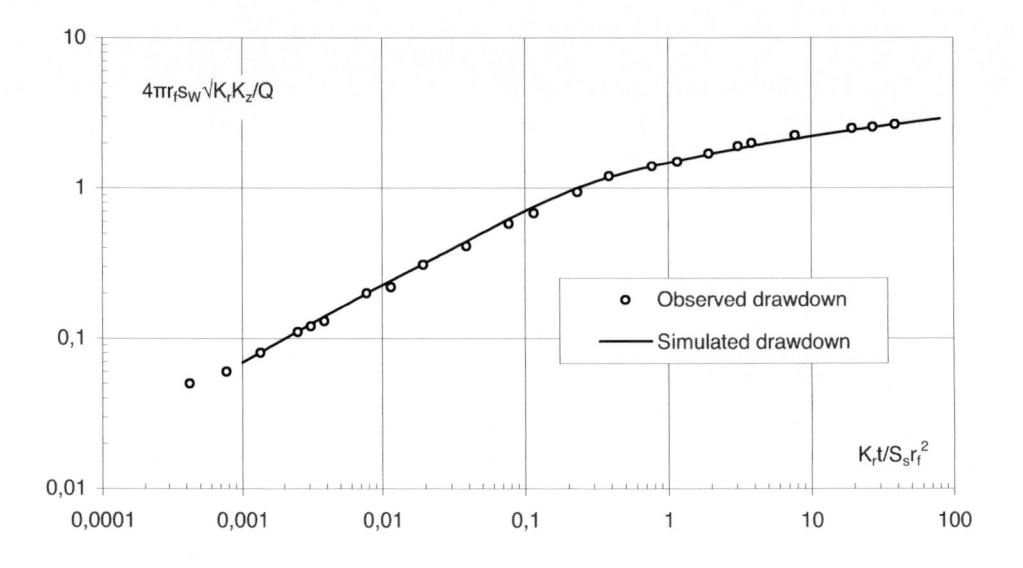

*Figure 8.7 Dimensionless log-log plot of drawdown and simulated drawdown with $h_D = 3$.*

## The straight-line method

Figure 8.8 shows drawdown data in a semi-log plot. We get the fracture transmissivity from Equation 8.8:

$$T = K_r h = \frac{0.183 Q}{\Delta s_w} = \frac{0.183 \cdot 95}{0.607 \cdot 3600} = 0.00796 \text{ m}^2/\text{s, or}$$

$$K_r = 0.000796 \text{ m/s}.$$

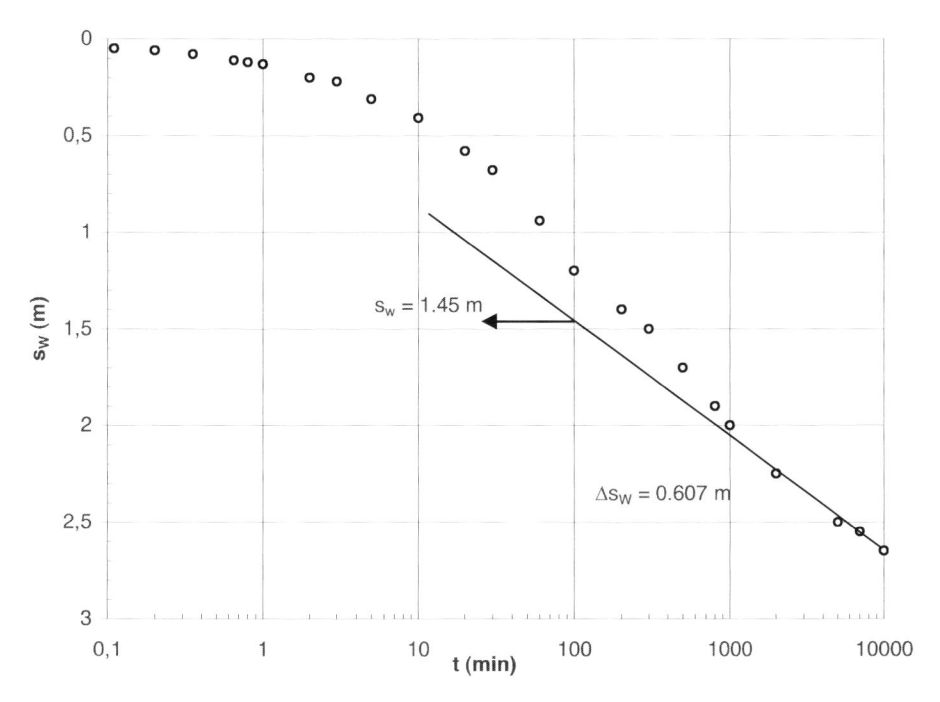

The skin factor, calculated from Equation 4.8, with $s_w = 1.45$ m for $t = 100$ min is

*Figure 8.8 Semi-log drawdown plot.*

$$\sigma = 1.1513 \left( \frac{s_w}{\Delta s_w} - \log t - \log \frac{T}{r_w^2 S} - 2.13 \right) =$$

$$1.1513 \left( \frac{1.45 - 0.8}{0.607} - \log 100 - \log \frac{0.00796}{0.1125^2 \cdot 0.00203} - 2.13 \right) = -6.39$$

Here we have subtracted the pseudo-skin loss.

The skin loss becomes

$$s_s = \frac{\sigma Q}{2\pi T} = -6.39 \frac{95}{2\pi \cdot 0.00796 \cdot 3600} = -3.37 \text{ m}.$$

The effective radius becomes

$$r_{ef} = r_w e^{-\sigma} = 0.1125 e^{6.39} = 67 \text{ m}.$$

The well efficiency at the end of the test is

$$V = \frac{s_w - s_s}{s_w} = \frac{2.65 + 3.37 - 0.8}{2.65} = 197\%$$

## 9. Diagnostics

Prior to data evaluation, it is important to carry out detailed data diagnostics to identify the physics behind the segments observed on drawdown curves. Without diagnostics, the wrong interpretation model may be selected and gross errors in evaluating aquifer parameter values may result. As we have seen, drawdown curves from wells in a fractured aquifer may behave very differently depending on the type of fractures present and the interaction between fractures and matrix. We can have several time periods characterizing the drawdown in the fractured well. A commonly seen error is to neglect the fact that the aquifer is fractured and consequently data is interpreted as if the aquifer material were sand. The changes in slope seen on semi-log plots are then often explained as the presence of leakage or barrier boundaries.

Well bore storage.

Well bore storage and/or fracture storage effects may influence early data. In a karstic aquifer, fracture storage may be considerably larger than well bore storage and behave in the same way. In wells with large yields, these effects may last up to a few minutes, often much shorter. The drawdown during this period is proportional to time. To identify well bore or fracture storage data, we plot on linear paper and early data should produce a straight line. If we plot data on log-log paper, early data produces a straight line with unit slope.

The naturally fractured aquifer

Drawdown curves in a naturally fractured or dual-porosity aquifer have a characteristic shape. In a semi-log plot the early data will fall on a straight line representing fractures alone. Late data will fall on a straight line parallel to the early line representing the homogeneous aquifer. A transition curve connects these two curve segments. The transition curve has a smaller log cycle slope representing matrix properties. Often only the late part of the transition curve and the late straight line appear in data from the pumped well. In this case, it may be difficult to obtain matrix parameters from pumped well data. If the early and late straight lines develop, the aquifer is definitely a dual-porosity aquifer with a well-developed fracture system. Data from monitoring wells placed near the pumped well will confirm this.

## Single fractures

In the sase of a *vertical fracture,* the drawdown curves may develop four distinct curve segments. First, we may see a very early period representing linear flow in the fracture. During this period, the drawdown increases with $\sqrt{t}$. On a log-log plot, this produces a straight line with slope 1:2. Plotted against $\sqrt{t}$ on linear paper, we get a straight line. Following the linear flow period, we may see a bilinear period, during which the drawdown is proportional to $\sqrt[4]{t}$. Data plotted against $\sqrt[4]{t}$ produces a straight line with slope 1:4 on log-log paper, and a straight line when plotted against $\sqrt[4]{t}$ on linear paper. The third curve segment represents linear formation flow, and this segment should always be seen, provided the test is long enough. Here, drawdown increases with $\sqrt{t}$. After a transition period, we see the late curve segment where the drawdown increases with $\log t$. This curve segment matches a Theis curve and produces a straight line on a semi-log plot.

In the case of a *horizontal fracture,* the drawdown curve may develop three distinct curve segments. First, we may see a short period with radial flow in the fracture. During this period, the drawdown increases with $\log t$. The second period shows vertical linear flow in the formation, and draw rown increases with $\sqrt{t}$. After a transition period we see pseudo radial flow. The drawdown increases with $\log t$. The horizontal fracture does not develop a period with bilinear flow.

In most cases type-curve matching on log-log paper will show whether the fracture is horizontal or vertical. The skin factor for these fractures will always be negative and the effective radius considerably larger than the well radius. The following examples demonstrate some diagnostics tools.

## Example 9.1 Well bore storage

Figure 9.1 shows synthetic data from a pumped well. It is more or less standard practice to use late data to determine transmissivity as shown. The drawdown curve is curvilinear all the way, which requires diagnostics. Figure 9.2 shows the same data in a linear plot. It becomes obvious that data shows only well bore storage and the test is not long enough to indicate aquifer behavior.

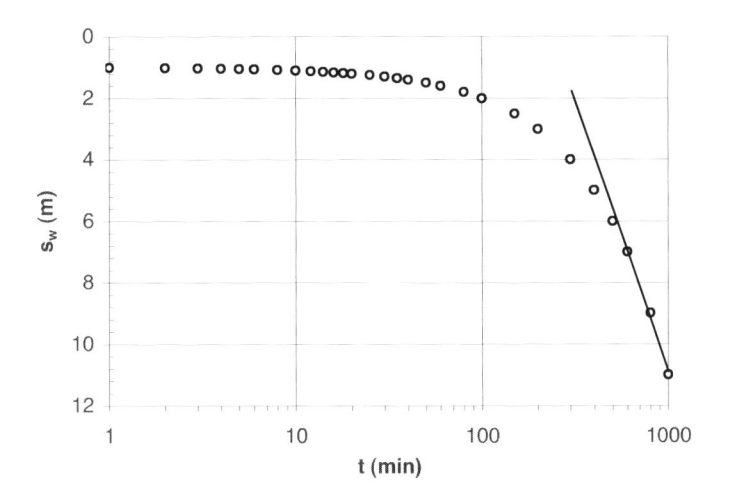

*Figure 9.1 Synthetic drawdown data from a pumped well in a semi-log plot.*

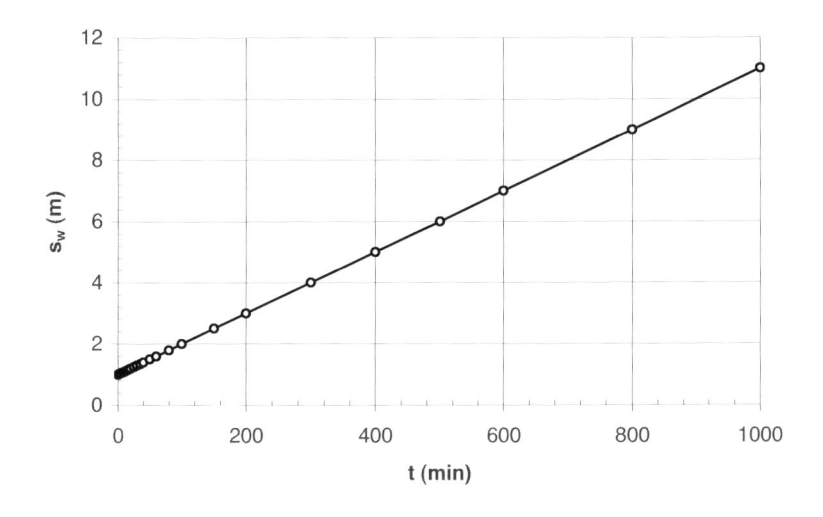

*Figure 9.2 Synthetic drawdown data in a linear plot.*

Example 9.2 Horizontal fracture

Data is from *Utah Department of Natural Resources, (2000)*. Figure 9.3 shows drawdown data from a pumped sandstone well in the Grapevine Pass Aquifer. The straight line shown was used to determine transmissivity, assuming that data up to about 600 min is influenced by well bore storage. To check this, we plot drawdown

143

# FRACTURED AQUIFERS

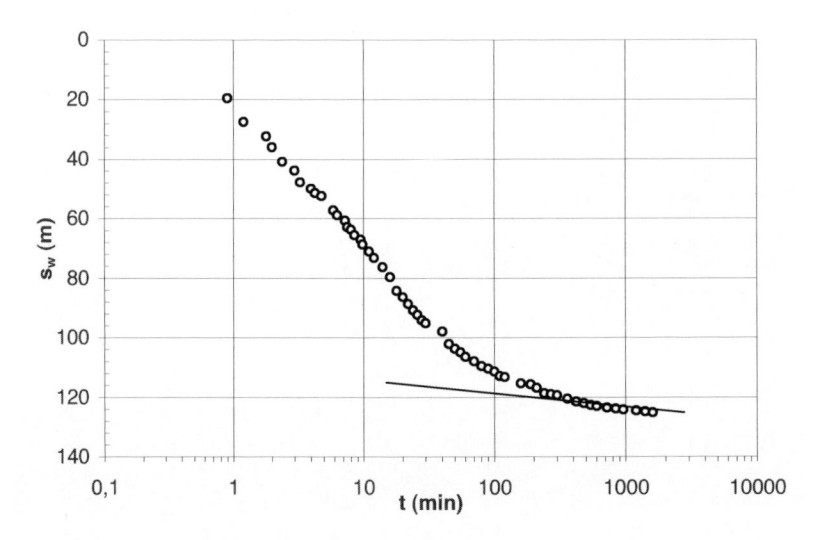

*Figure 9.3 Semi-log plot of drawdown data from a sandstone well in the Grapevine Pass Aquifer.*

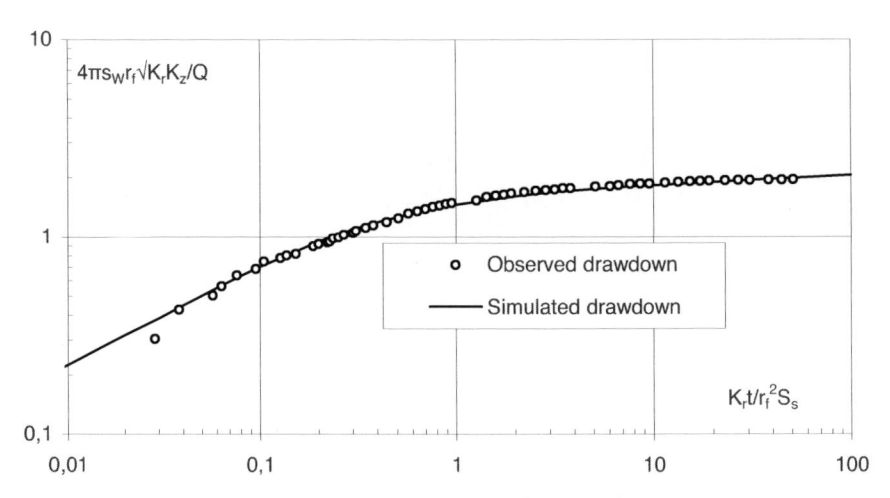

*Figure 9.4 Log-log plot of drawdown data from figure 9.3 with simulated drawdown assuming a horizontal single fracture with $h_D = 10$.*

data on log-log paper, Figure 9.4. This should produce a straight line with unit slope. We get a straight line with slope 1:2 indicating a single fracture. Well bore storage influences only the first 2 - 3 data points. Actually, drawdown data matches the type curve for a single horizontal fracture with $h_D = 10$.

144

Example 9.3

Data is again from *Utah Department of Natural Resources (2000)*. Figure 9.6 shows drawdown data from a monitoring well in a semi-log plot. The drawdown is curvilinear and it may not be correct to use the straight line indicated for parameter evaluation unless justified by diagnostics. Figure 9.7 shows the same data plotted on log-log paper and against the square root of time in Figure 9.8.

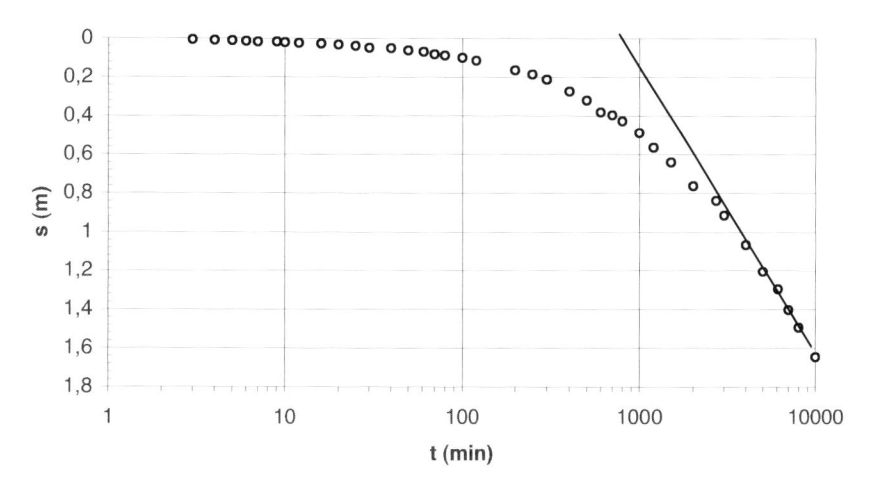

*Figure 9.6 Drawdown data from a monitoring well in the New Harmony Aquifer.*

Figure 9.7 indicates that the drawdown curve does not match a Theis curve. Late data forms a line with slope = ½. Figure 9.8 shows that the drawdown increases with $\sqrt{t}$. Although the test duration is a week, it is not long enough to reveal homogeneous aquifer behavior. Data shows that the monitoring well is located in a single fracture. It is not possible to determine whether the fracture is vertical or horizontal. Data from the pumped well would probably resolve this.

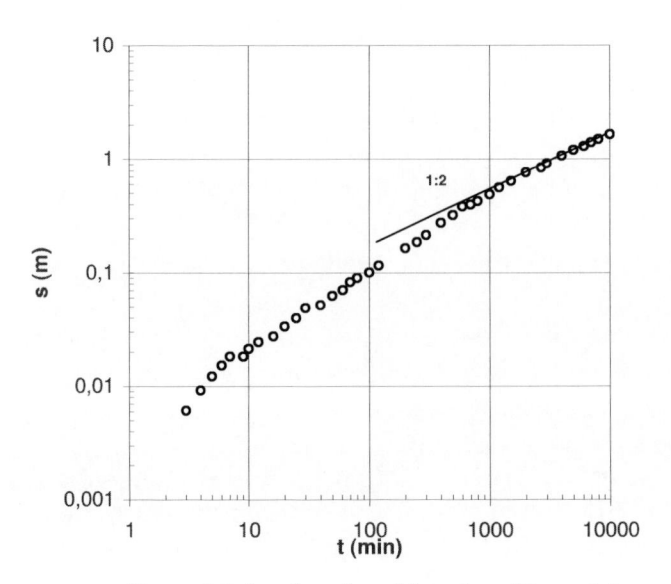

*Figure 9.7. Log-log plot of data from Figure 9.6*

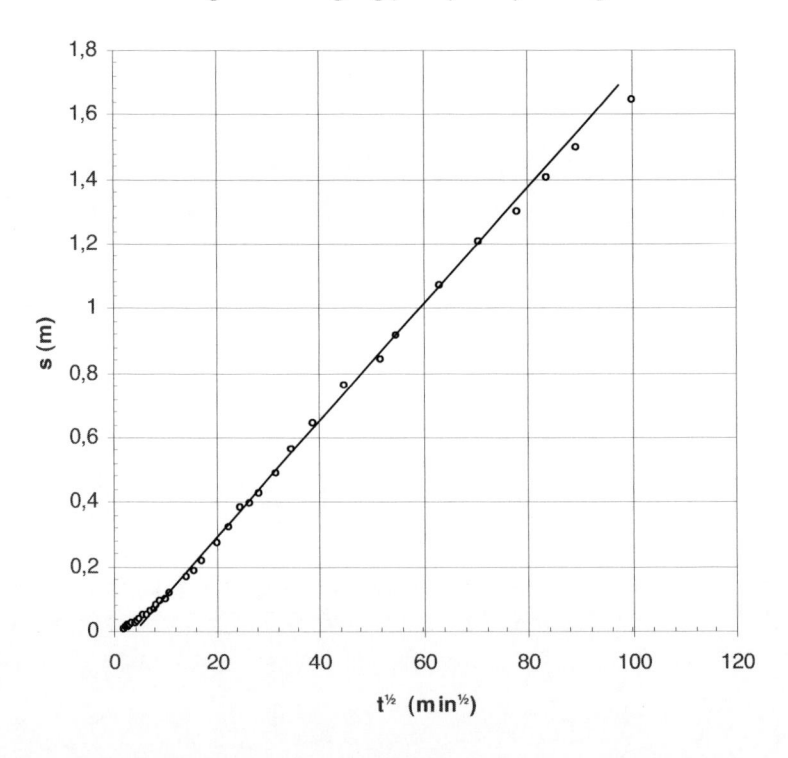

*Figure 9.8 Drawdown data from Figure 9.6 plotted against the square root of time.*

## 10. Brief guidelines to fractured aquifer testing

The purpose of testing fractured wells is to obtain data, which provide information about fractures and matrix. Early data reveals fracture properties, late data homogeneous aquifer behavior and the transition period gives information about matrix properties. To obtain the data, the test should meet some basic requirements.

<u>The pumped well</u>

Data from the pumped well is the key to understanding the nature of the fractures. It is imperative that the early monitoring be frequent, like one measurement every 10 seconds for the first few minutes. After this, one measurement per minute is sufficient up to 20 minutes and then the frequency can be gradually reduced. The use of data loggers is necessary.

Well yield should be as high as possible without dewatering the aquifer. This will reduce the relative influence from changes in atmospheric pressure and possible other pumping. It would be an advantage to carry out a short test of 10 – 20 minutes duration to get an idea of the magnitude of the drawdown during the constant discharge test. If the aquifer is confined a data logger should be used to monitor changes in atmospheric pressure. Corrections for this could be necessary, particularly for monitoring well data. It is recommended to install data loggers in the pumped well and in monitoring wells at least a few days prior to the test. This can provide data for determination of the barometric efficiency and give information about the possible influence of other pumping from the aquifer.

The test duration required is difficult to determine beforehand. An indispensable requirement is that pumped well data should reveal homogeneous aquifer behavior and, if possible, monitoring well data as well. In a naturally fractured aquifer, a duration of one week is in many cases sufficient. For aquifers with single fractures, the test may have to last for weeks before data sufficient for fracture type identification is available. The test should continue until preliminary data analysis shows that sufficient data is at hand.

The discharge should be kept constant. All methods of interpretation assume a constant discharge rate. Frequent monitoring of the discharge rate is imperative.

<u>Monitoring wells</u>

Monitoring well data is necessary for storage evaluation. In confined aquifers, the cone of depression develops rapidly and frequent monitoring (every minute) early on during the test is recommended. Later during the test, after 20 minutes, 5 – 10 minutes intervals may be used.

If possible, the test should last until monitoring well data shows the homogeneous aquifer. This will usually happen later than in the pumped well in a dual-porosity system. For single fractures, monitoring wells may display homogeneous behavior earlier than the pumped well.

Monitoring wells located a large distance from the pumped well (600 – 800 m or more) may display the homogeneous aquifer from the beginning. To obtain information on matrix it is necessary to include monitoring wells within a short distance, say, wells less than 100 m away. Fractured aquifers are usually strongly anisotropic. It is a great advantage to monitor wells in various directions and distances.

## 11. Variable discharge

All interpretation methods described in previous chapters assume a constant discharge rate. If the discharge rate is not constant, these methods do not apply. In cases where the discharge rate fluctuates around a constant mean value, the methods still apply. However, the data points used in the analysis must be taken when the discharge rate is at its mean value. Correcting data for changes in discharge rate is uncertain, particular for high-yielding wells, because turbulent losses may be large and often unknown.

If the discharge rate decreases steadily, drawdown data may erroneously indicate leakage or dual-porosity effects. If the discharge rate increases steadily, data may erroneously indicate the influence of a barrier boundary. Using a mean value of the discharge rate in such cases may result in gross errors.

*Abu-Zied and Scott (1963)* solved the case of exponentially decreasing discharge in a confined aquifer. *Hantush (1964a, 1964b)* solved the cases of exponentially and hyperbolically decreasing discharge in a confined aquifer and in a leaky system without storage in the semi-pervious layers. *Sen and Altunkaynak (2005)* also considered the case of variable discharge rate and devised a straight-line method for analysing late data for exponentially decreasing discharge.

While the variable discharge case has limited application for well testing, it is particularly useful for analysing historical well field data. For well fields the discharge rate has rarely been constant because of variations in water demand. Water supplies often have records of discharge and drawdowns dating many decades back. A history match with such data can provide useful information about the aquifers in the long term. The advantage of an analytical approach is that the long-term hydraulic parameters determined from a history match include the influence of inhomogeneities and hydraulic barriers. Because of the long-term development of the aquifer investigated, historical data reveals homogeneous aquifer behavior. The only parameters we need to know to describe drawdowns in the aquifer are the long-term values of aquifer transmissivity and storage coefficient. The analysis is applicable to fractured aquifers as well as multilayer aquifer systems.

The usual application of historical data is to use it as input for calibration of digital ground water models. Such models have limited application for long-term simulation (several decades) because more or less incorrect model boundaries may result in incorrect future drawdowns. The use of analytical models eliminates this problem because boundary effects are included in the model. Analytical models are much simpler and faster to use than digital models and sometimes analytical models produce results that are more accurate.

The basic assumption behind the analytical model is that the aquifer behaves homogeneously. The long-term aquifer transmissivity is that of the pumped layer and the storage coefficient is the sum of storage coefficients in each layer. Since the water-table aquifer on top of the layers has a storage coefficient that is orders of magnitudes larger than confined storage coefficients, the aquifer system exhibits phreatic behavior in the long term.

The method described here applies the type-curve procedure. To produce type curves, we need to find the equation for the drawdown using the actual discharge function. One method is to solve the governing differential equation with the discharge function as a boundary condition. A more direct method is to integrate point sources. *Carslaw and Jaeger (1959)* describe the procedure in detail and develop solutions to the point and line source, plane sources and surface sources. Here, we are interested in line-source solutions. Once found, it is simple to plot the solutions as type curves on log-log paper and apply the Match Point procedure to determine hydraulic parameters from the history match.

*11.1 The continuous line source*
We consider a fully penetrating well pumping from time $t = 0$ at the rate $Qf(t)$ in an infinite homogeneous and isotropic aquifer. The solution for the drawdown at time $t$ is *(Carslaw and Jaeger, 1959)*

$$s = \frac{Q}{4\pi T} \int_0^t f(t') \exp\left(-r^2 S / 4T(t-t')\right) \frac{dt'}{t-t'} \qquad (11.1)$$

We introduce the substitutions

$$u = \frac{r^2 S}{4Tt}, \qquad x = \frac{r^2 S}{4T(t-t')}$$

and obtain

$$t = \frac{r^2 S}{4T} \cdot \frac{1}{u}, \qquad t' = t - \frac{r^2 S}{4Tx} = \frac{r^2 S}{4T}\left(\frac{1}{u} - \frac{1}{x}\right)$$

$$dx = \frac{r^2 S}{4T(t-t')} dt', \quad \frac{dx}{x} = \frac{dt'}{t-t'}$$

The integration limits become

$$t' = 0 \Rightarrow x = u \text{ and } t' = t \Rightarrow x = \infty$$

Equation 11.1 now reads

$$s = \frac{Q}{4\pi T} \int_u^\infty f\left(\frac{r^2 S}{4T}\left(\frac{1}{u} - \frac{1}{x}\right)\right) \frac{e^{-x}}{x} dx \qquad (11.2)$$

For the special case, $f(t) = 1$ we obtain Theis' solution

$$s = \frac{Q}{4\pi T} \int_u^\infty \frac{e^{-x}}{x} dx$$

To find solutions to field problems we need to determine the function $f(t)$ empirically from data and substitute into Equation 11.2. The function must be relatively simple in order to obtain solutions containing tabulated integrals. If not, numerical integration will be necessary. Here, we consider linear, exponential and hyperbolic functions. In carrying out the integrations, the following equations are useful (*Gradshteyn and Ryzhik, 1965*),

$$\int_u^\infty \frac{e^{-x}}{x^2} dx = -W(u) - \frac{e^{-u}}{u}$$

$$\int_u^\infty \frac{e^{-x}}{x-c} dx = e^{-c} W(u-c)$$

$$\int_u^\infty \frac{1}{x} \exp\left(-x - \frac{\beta^2}{4x}\right) dx = W(u, \beta)$$

The last integral is the well function for leaky aquifers. *Hantush (1956)* evaluated the integral in tabular form.

# FRACTURED AQUIFERS

## 11.2 Exponentially varying discharge
### 11.2.1 Exponential decrease

The discharge function chosen is

$$f(t) = Q_2 + (Q_1 - Q_2)e^{-kt} = Q_2\left(1 + (\alpha - 1)e^{-kt}\right) \qquad (11.3)$$

where $Q_2$ is the final discharge, $Q_1$ is the initial discharge, $\alpha = Q_1/Q_2 > 1$ and $k$ is a time constant, see Figure 11.1. These parameters must be determined empirically from data.

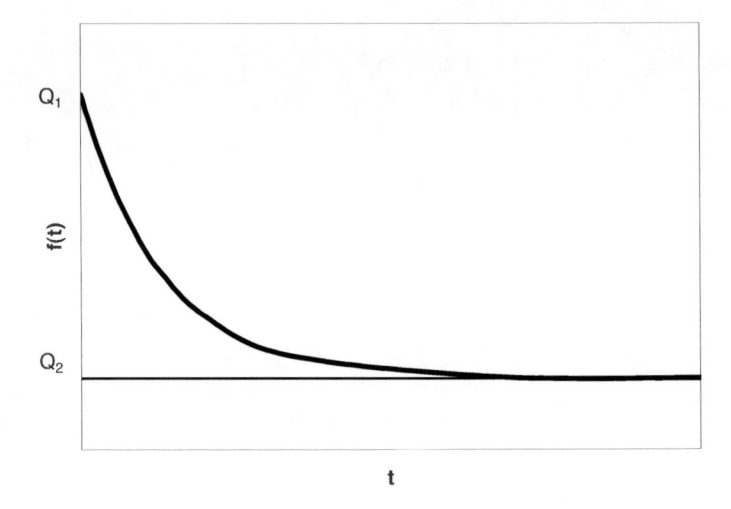

*Figure 11.1 Exponentially decreasing discharge.*

Inserting Equation 11.3 into equation 11.2 yields

$$s = \frac{Q_2}{4\pi T} \int_u^\infty \left(1 + (\alpha - 1)\exp\left(-\frac{kr^2 S}{4T}\left(\frac{1}{u} - \frac{1}{x}\right)\right)\frac{e^{-x}}{x}\right) dx \qquad (11.4)$$

With $\beta^2 = kr^2 S/T$, Equation 11.4 becomes

$$s = \frac{Q_2}{4\pi T}\left(\int_u^\infty \frac{e^{-x}}{x}dx + (\alpha-1)\exp(-\beta^2/4u)\int_u^\infty \exp(-x-\beta^2/4x)\frac{dx}{x}\right) =$$
$$\frac{Q_2}{4\pi T}\left(W(u) + (\alpha-1)\exp(-\beta^2/4u)W(u,\beta)\right)$$

(11.5)

Equation 11.5 is essentially the solution found by *Sen and Altunkaynak (2004)* and *Hantush (1964)*. Figure 11.2 shows type curves based on Equation 11.5 for $\alpha = 5$. The equations for determination of transmissivity and storage are

$$T = \frac{Q_2}{4\pi s_{MP}} \quad \text{and} \quad S = \frac{4Tt_{MP}}{r^2}$$

The parameter $\beta$ does not provide any new information since we know the values of $k, r, T$ and $S$.

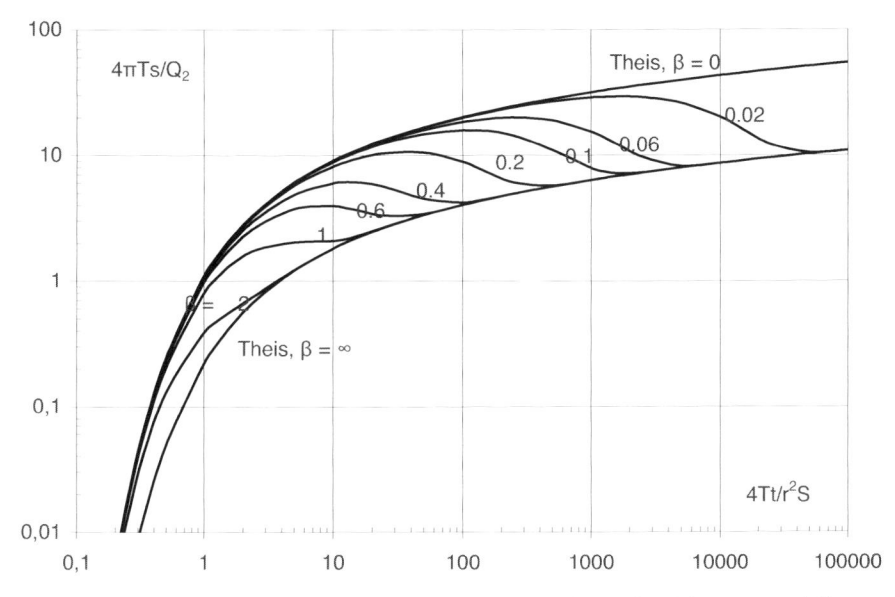

Figure 11.2 Type curves for drawdown around a well with exponentially decreasing discharge, $\alpha = 5$.

Example 11.1

Data is from a well field discharged by a siphon system. In such a system, wells produce water under a constant vacuum resulting in a decrease in discharge with time. The pumped aquifer is a confined fractured limestone aquifer at Havdrup

FRACTURED AQUIFERS

south of Copenhagen, operated by Copenhagen Energy. The test duration is long enough to reveal homogeneous aquifer behavior. Data in the table shows the discharge rate, drawdown and recovery in a monitoring well 600 m from the well field center.

The discharge function chosen is

$$f(t) = 9.000(1 + 0.75\exp(-0.8t))$$

that is, $\alpha = 1.75$.

Figure 11.4 shows drawdown and recovery data from the monitoring well. Drawdown data shows the influence of decreasing discharge while recovery data follow a Theis curve subject to the final discharge rate. Taking a mean value of the discharge rate will obviously give misleading results..

| t (days) | Q (m³/d) | t (days) | s (m) | Δt (days) | s-s' (m) |
|---|---|---|---|---|---|
| 1 | 11600 | 0,43 | 0.8 | 0.28 | 0.29 |
| 1.5 | 10600 | 1 | 1.7 | 0.56 | 0.58 |
| 2 | 9650 | 19 | 2.4 | 0.83 | 0.78 |
| 3 | 9800 | 2.9 | 2.7 | 1.5 | 1.15 |
| 4 | 9550 | 4.9 | 2.9 | 1.9 | 1.3 |
| 5 | 9300 | 5.8 | 3 | 2.9 | 1.6 |
| 6 | 9220 | 6.8 | 3 | 3.9 | 1.8 |
| 7 | 9200 | 7.7 | 3 | 4.8 | 2.1 |
| 8 | 9100 | 8.8 | 3.1 | 5.8 | 2.2 |
| 9 | 9180 | 10.5 | 3.1 | 7.8 | 2.5 |
| 10 | 8950 | 13.5 | 3.1 | 9.5 | 2.6 |
| 11 | 8950 | 15.5 | 3.2 | 12 | 2.7 |
| 12 | 8950 | 18 | 3.3 | 13.5 | 2.8 |
| 13 | 8950 | | | 15.5 | 2.9 |
| 14 | 9090 | | | 18 | 3.2 |
| 15 | 9090 | | | | |
| 16 | 9090 | | | | |
| 17 | 8950 | | | | |
| 18 | 9070 | | | | |

Figure 11.3 shows the discharge rate as a function of time in days.

154

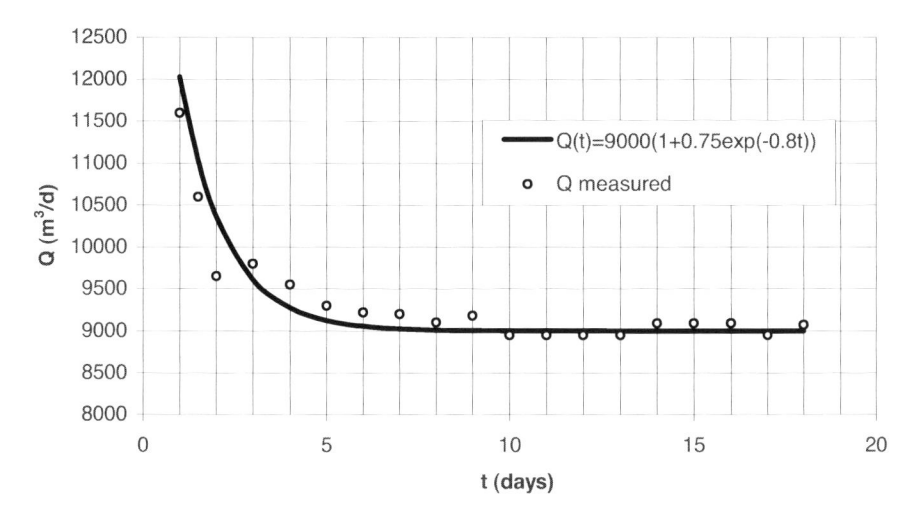

*Figure 11.3 Measured discharge rate.*

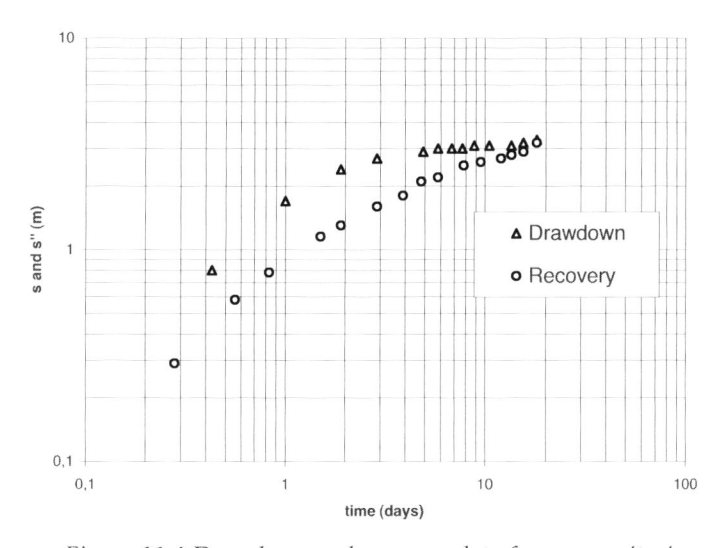

*Figure 11.4 Drawdown and recovery data from a monitoring well.*

Figure 11.4 shows a history match with variable discharge type curves for $\alpha = 1.75$.

# FRACTURED AQUIFERS

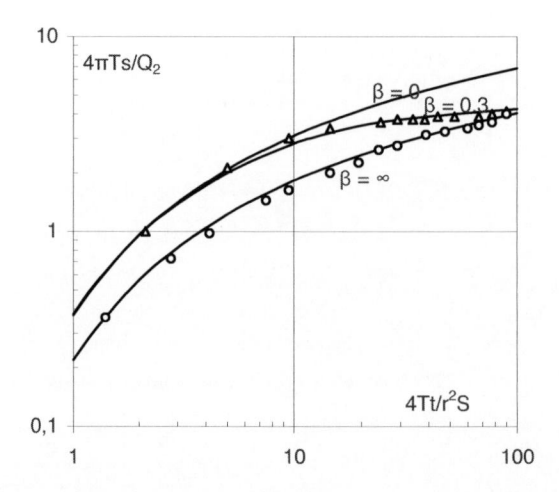

*Figure 11.5 History match with type curve, $\alpha = 1.75, \beta = 0.3$.*

## 11.2.2 Exponential increase

In this case, we choose the discharge function

$$f(t) = Q_2 + (Q_1 - Q_2)\exp(-kt) = Q_2(1 + (\alpha - 1)\exp(-kt)) \qquad (11.6)$$

where $\alpha = Q_1/Q_2 < 1$. Figure 11.6 shows a schematic of the function.

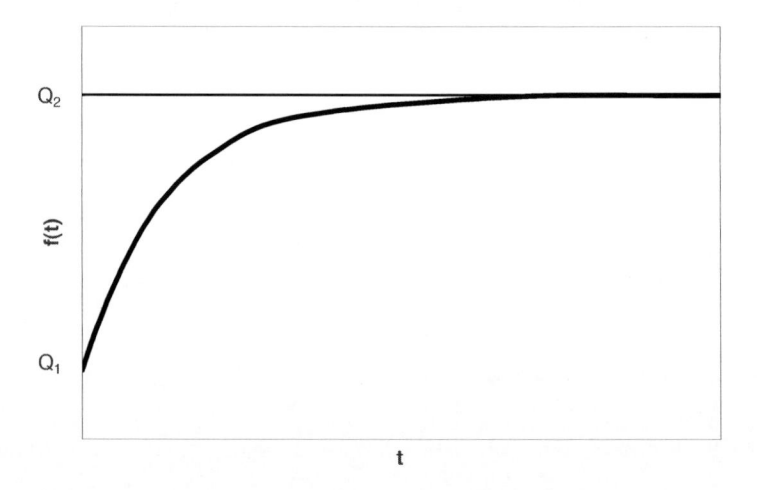

*Figure 11.6 Exponentially increasing discharge.*

Substituting this equation into Equation 11.2 gives

$$s = \frac{Q_2}{4\pi T}\left(W(u) + (\alpha - 1)\exp(-\beta^2/4u)W(u,\beta)\right) \tag{11.7}$$

Figure 11.7 shows the type-curve plot based on equation 11.7.

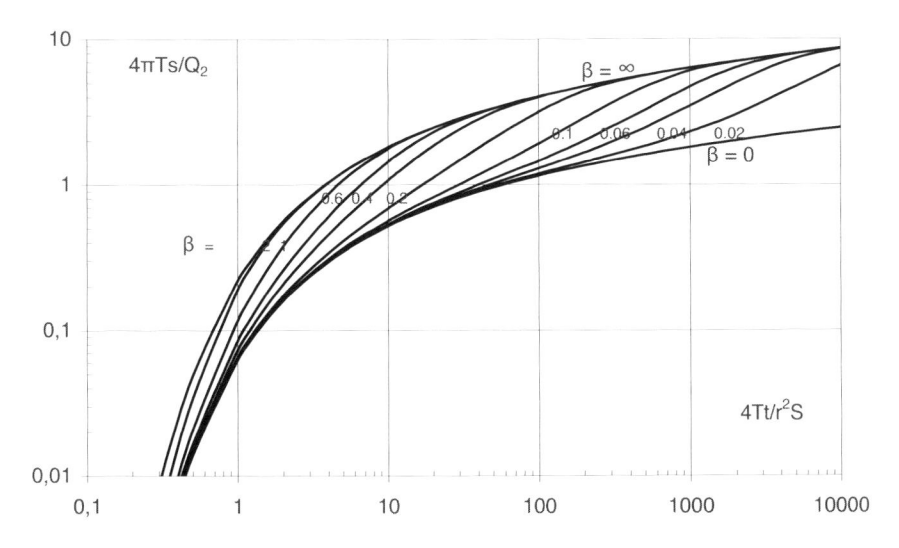

*Figure 11.7 Type curves for exponentially increasing drawdown, $1/\alpha = 3.5$.*

Example 11.2

Data is from the Holmehave well field operated by the Municipal Water Supply of Odense, Denmark *(Nielsen, 2006 a)*. The well field started pumping in 1978, and the wells tap two confined sand aquifers separated by clay layers. Monthly data on water levels and discharge is available. Figure 11.8 shows the total discharge as a function of time.

The initial discharge rate is $Q_1 = 120,000$ m³/month, the final discharge rate is $Q_2 = 420,000$ m³/month and $1/\alpha = 3.5$. The discharge function, Equation 11.6, with actual values is

$$f(t) = 420,000(1 - 0.714\exp(-0.0196t))$$

It appears from Figure 11.8 that this function is valid up to about 210 months.

# FRACTURED AQUIFERS

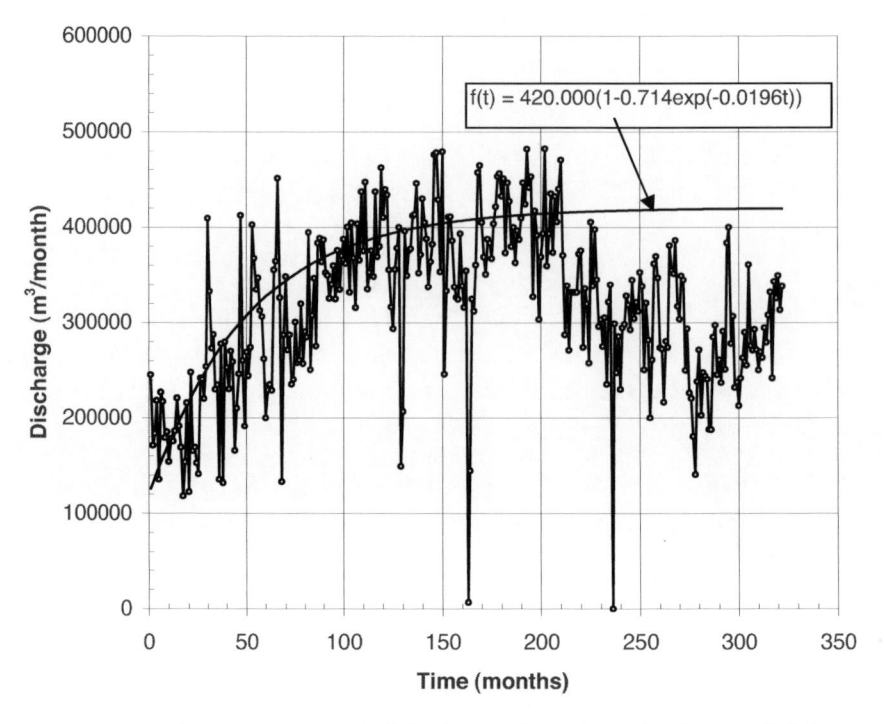

*Figure 11.8 Historical discharges from the Holmehave well field.*

Figure 11.9 shows drawdown data from a monitoring well, located 1185 m from the well field centre. The table presents data. Matching with type curves, Figure 11.7, gives $s_{MP} = 1.1$, $t_{MP} = 11.5$ days and $\beta = 0.1$.

The transmissivity is

$$T = \frac{Q_2}{4\pi s_{MP}} = \frac{420.000}{4\pi \cdot 1.1 \cdot 30.4 \cdot 24 \cdot 3600} = 0.0116 \text{ m}^2/\text{s}$$

The storage coefficient is

$$S = \frac{4Tt_{MP}}{r^2} = \frac{4 \cdot 0.0116 \cdot 11.5 \cdot 1440 \cdot 60}{1185^2} = 0.0328$$

| t (days) | s (m) | t (days) | s (m) |
|---|---|---|---|
| 13 | 0.16 | 1787 | 2.6 |
| 72 | 0.42 | 2399 | 2.97 |
| 100 | 1.13 | 2491 | 2.97 |
| 334 | 1.19 | 2613 | 3.7 |
| 483 | 1.25 | 2794 | 3.58 |
| 636 | 1.33 | 3008 | 3.33 |
| 847 | 1.61 | 3374 | 3.94 |
| 1122 | 1.81 | 3652 | 3.99 |
| 1370 | 2.66 | 3864 | 4.1 |
| 1577 | 2.81 | 4142 | 4.27 |
|  |  | 4408 | 4.49 |

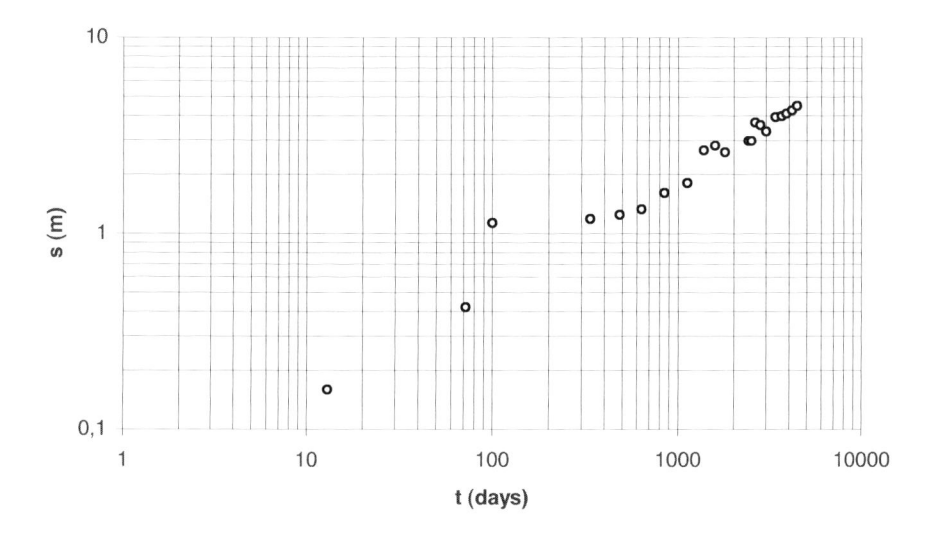

*Figure 11.9 Drawdown data from a monitoring well, $r = 1185$ m.*

Figure 11.10 shows the type-curve match in a dimensionless plot.

We note that the storage coefficient is in the water-table range although the aquifer system is confined. This is a general observation in well-developed aquifers.

### *11.3 Hyperbolically varying discharge*
### 11.3.1 Hyperbolic decrease

The discharge function chosen is

# FRACTURED AQUIFERS

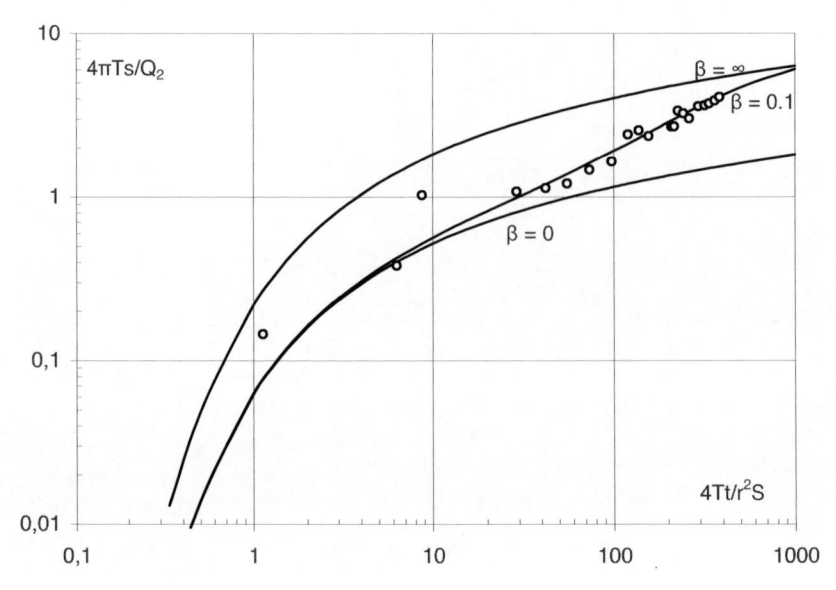

*Figure 11.10 The Holmehave well field. Type-curve match with drawdown data from a monitoring well.*

$$f(t) = Q_2 + \frac{Q_1 - Q_2}{1 + kt} = Q_2\left(1 + \frac{\alpha - 1}{1 + kt}\right), \quad \alpha = Q_1/Q_2 > 1 \tag{11.8}$$

Figure 11.11 shows a schematic of this discharge function.

Substituting Equation 11.8 into Equation 11.2 gives

$$s = \frac{Q_2}{4\pi T} \int_u^\infty \left(1 + \frac{\alpha - 1}{\frac{kr^2 S}{4T}(1/u - 1/x) + 1}\right) \frac{e^{-x}}{x} dx =$$

$$\frac{Q_2}{4\pi T}\left(\int_u^\infty \frac{e^{-x}}{x} dx + (\alpha - 1)\int_u^\infty \frac{e^{-x}}{x(\beta^2/4u + 1) - \beta^2/4} dx\right) =$$

$$\frac{Q_2}{4\pi T}\left(W(u) + \frac{\alpha - 1}{1 + \beta^2/4u}\int_u^\infty \frac{e^{-x}}{x - \beta^2/4(1 + \beta^2/4u)} dx\right) = \tag{11.9}$$

160

$$\frac{Q_2}{4\pi T}\left(W(u) + \frac{(\alpha-1)\exp\left(-1/\left(4/\beta^2 + 1/u\right)\right)}{1+\beta^2/4u} W\left(u - \frac{1}{4/\beta^2 + 1/u}\right)\right)$$

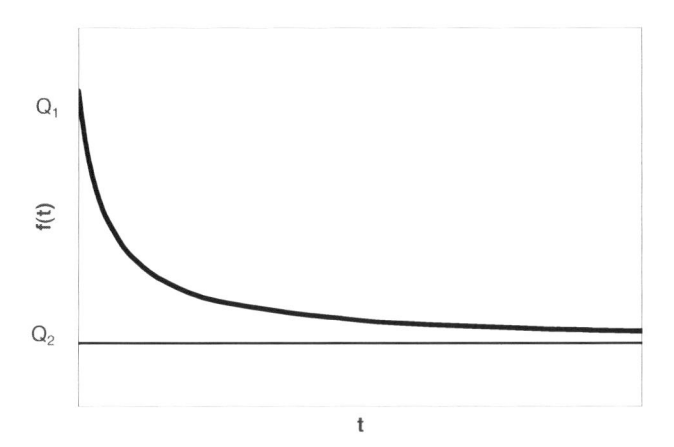

*Figure 11.11 Hyperbolically decreasing discharge.*

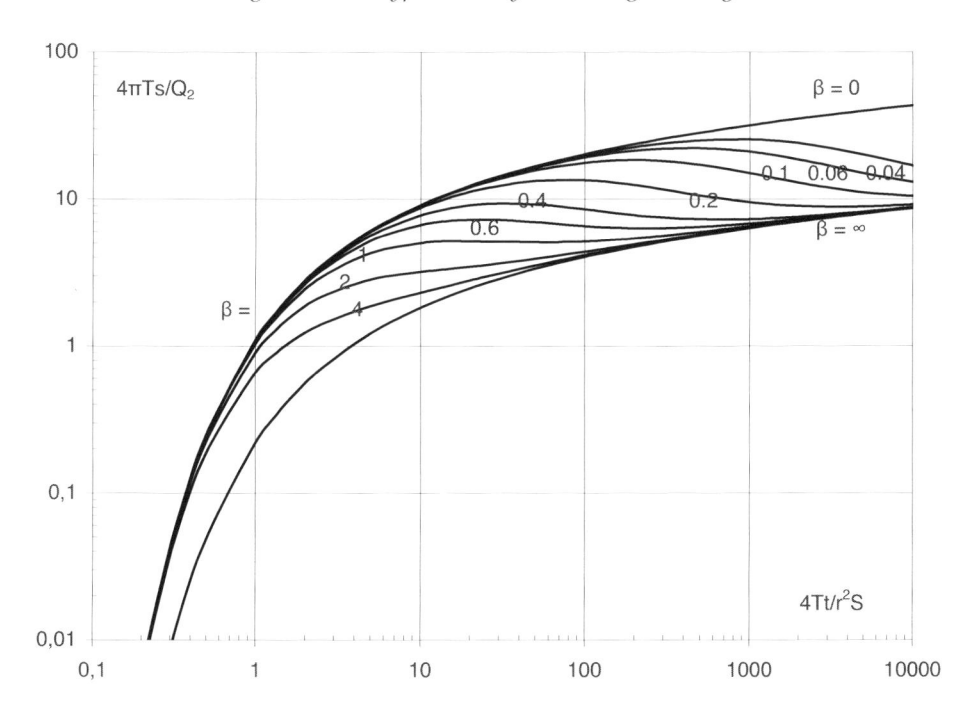

*Figure 11.12 Type curves for hyperbolically decreasing discharge, $\alpha = 5$.*

Figure 11.12 shows type curves based on Equation 11.9.

## 11.3.2 Hyperbolic increase

The discharge function is as for hyperbolic decrease

$$f(t) = Q_2 + \frac{Q_1 - Q_2}{1 + kt} = Q_2\left(1 + \frac{\alpha - 1}{1 + kt}\right), \quad \alpha = Q_1/Q_2 < 1$$

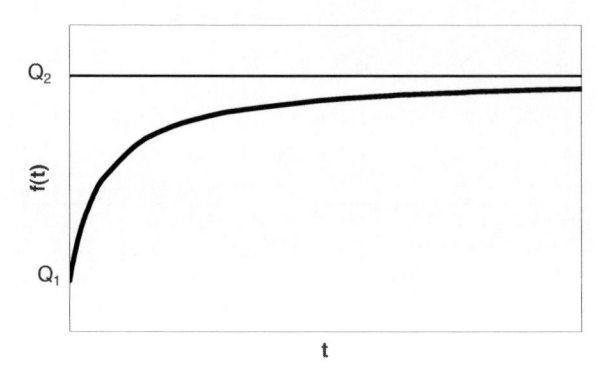

*Figure 11.13 Hyperbolically increasing discharge.*

and the equation for drawdown is

$$s = \frac{Q_2}{4\pi T}\left(W(u) + \frac{(\alpha - 1)\exp\left(-1/(4/\beta^2 + 1/u)\right)}{1 + \beta^2/4u} W\left(u - \frac{1}{4/\beta^2 + 1/u}\right)\right) \quad (11.10)$$

Figure 11.14 shows the type-curve plot for $\alpha = 0.2$.

### 11.4 Linear change in discharge
In this case, we consider a linear discharge increase. The discharge equation is

$$f(t) = Q_1(1 + kt) \tag{11.11}$$

Figure 11.15 shows the discharge function.

Substituting Equation 11.11 into Equation 11.2 gives

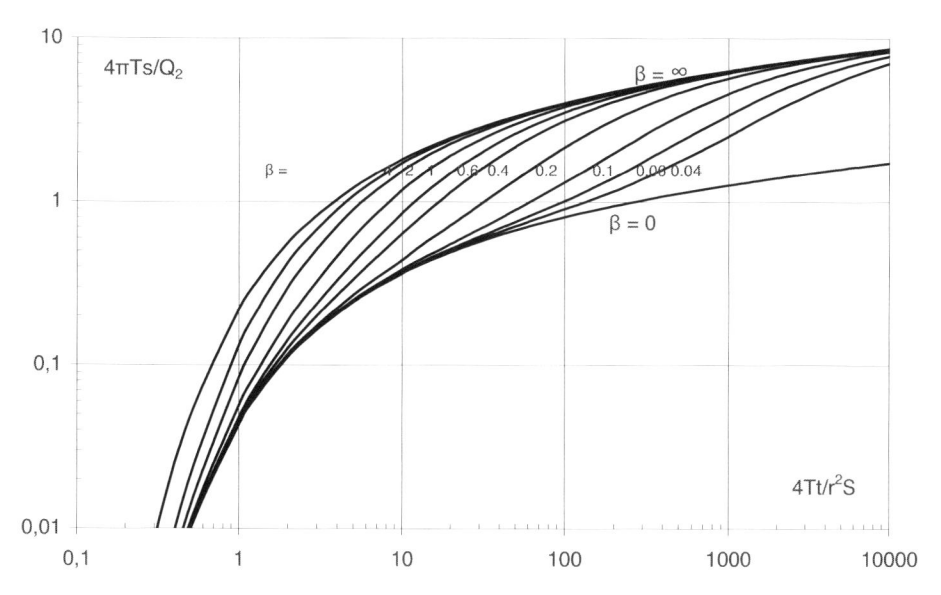

*Figure 11.14 Type curves for hyperbolically increasing discharge, $\alpha = 0.2$.*

$$s = \frac{Q_1}{4\pi T} \int_u^\infty \left(1 + \frac{kr^2 S}{4T}\left(\frac{1}{u} - \frac{1}{x}\right)\right) \frac{e^{-x}}{x} dx =$$

$$\frac{Q_1}{4\pi T} \left( \int_u^\infty \frac{e^{-x}}{x} dx + \frac{\beta^2}{4} \int_u^\infty \frac{e^{-x}}{ux} dx - \frac{\beta^2}{4} \int_u^\infty \frac{e^{-x}}{x^2} dx \right)$$

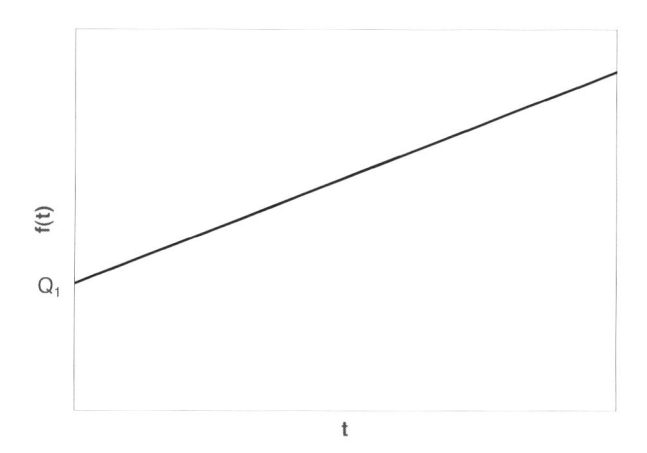

*Figure 11.15 Linear increase in discharge.*

FRACTURED AQUIFERS

The final solution for the drawdown becomes

$$s = \frac{Q_1}{4\pi T}\left(\left(1+\frac{\beta^2}{4}\left(1+\frac{1}{u}\right)\right)W(u)+\frac{\beta^2}{4}\cdot\frac{e^{-u}}{u}\right) \qquad (11.12)$$

Figure 11.16 shows the type-curve plot. For large values of time (small values of $u$) the drawdown curve becomes a straight line with unit slope.

A special case is

$$f(t) = Q_0 t \qquad (11.13)$$

i.e. zero initial discharge. Here, $Q_0$ is the slope of the discharge curve (dimension $L^3/T^2$). Substituting Equation 11.13 into Equation 11.2 gives

$$s = \frac{Q_0 r^2 S}{16\pi T^2}\int_u^\infty\left(\frac{1}{u}-\frac{1}{x}\right)\frac{e^{-x}}{x}dx = \frac{Q_0 r^2 S}{16\pi T^2}\left(\left(1+\frac{1}{u}\right)W(u)+\frac{e^{-u}}{u}\right) \qquad (11.14)$$

Example 11.3

Data is from the Beder well field operated by the municipal water works of Aarhus, Denmark, *(Houlberg et al, 1993)*. The aquifer is a confined sand aquifer overlain by 40-80 m of boulder clay. Figure 11.17 shows discharge data as a function of time.

The discharge data is approximately a straight line with initial discharge rate $Q_1 = 200,000$ m³/month. Figure 11.18 shows drawdown data from a monitoring well at distance 948 m from the well field centre. The table shows drawdown data.

Matching drawdown data with the type curves, Figure 11.16, gives $s_{MP} = 0.5$ m, $t_{MP} = 0.33$ days and $\beta = 0.02$. We get the transmissivity

$$T = \frac{Q_1}{4\pi T} = \frac{200.000}{4\pi\cdot 30.4\cdot 24\cdot 3600} = 0.0121 \text{ m}^2/\text{s}$$

The storage coefficient becomes

$$S = \frac{4Tt_{MP}}{r^2} = \frac{4 \cdot 0.0121 \cdot 0.33 \cdot 1440 \cdot 60}{948^2} = 0.0015$$

Figure 11.19 shows the data match.

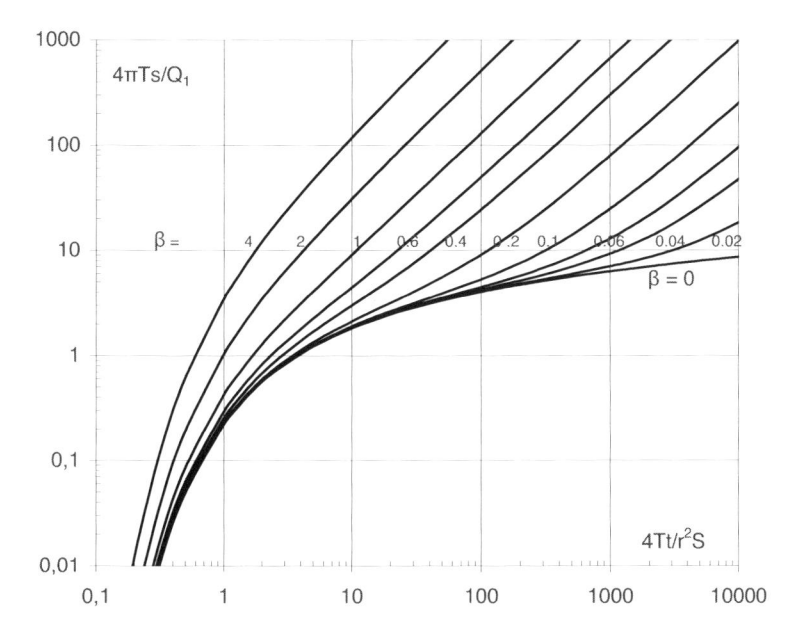

*Figure 11.16 Type curves for linear discharge increase.*

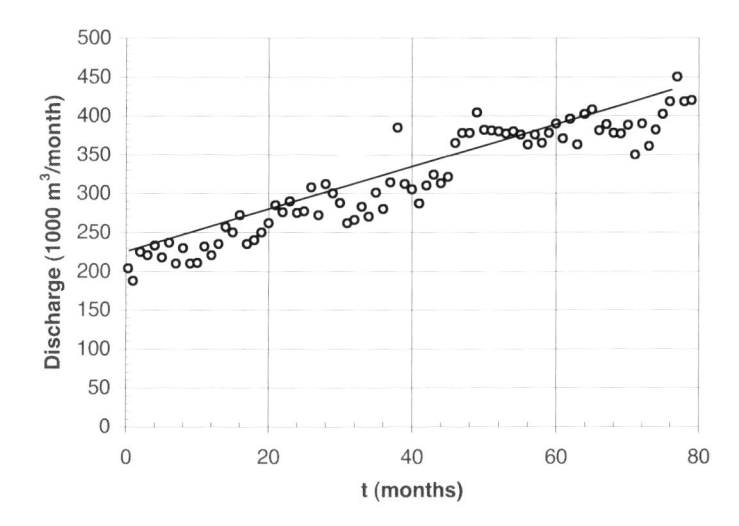

*Figure 11.17 The Beder well field. Monthly discharge for the first 79 months of operation.*

# FRACTURED AQUIFERS

| t (days) | s (m) | t (days) | s (m) |
|---|---|---|---|
| 59 | 2.85 | 995 | 5.1 |
| 160 | 2.99 | 1020 | 5.1 |
| 255 | 2.7 | 1200 | 4.98 |
| 330 | 3.08 | 1400 | 5 |
| 440 | 2.7 | 1600 | 5.5 |
| 500 | 4 | 1800 | 5.7 |
| 560 | 4.09 | 2000 | 6.55 |
| 620 | 4.45 | 2100 | 6.91 |
| 725 | 4.5 | 2300 | 7.5 |
| 820 | 4.91 | 2600 | 7.75 |
| 870 | 4.8 | 2750 | 8.85 |

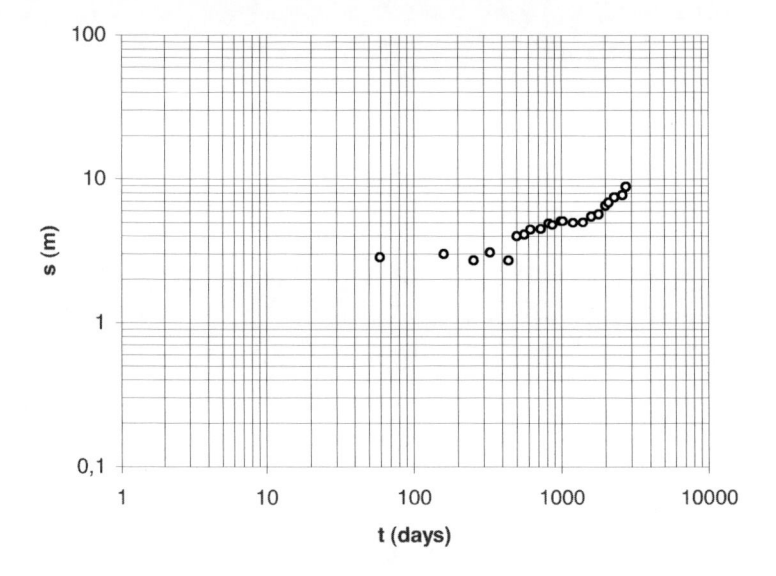

*Figure 11.18. The Beder well field, drawdown data from a monitoring well.*

In this case, we do not get a storage value in the water-table range. The reason is probably that the discharge increases continually and the aquifer never settles in its long-term stage but reveals only short-term behavior.

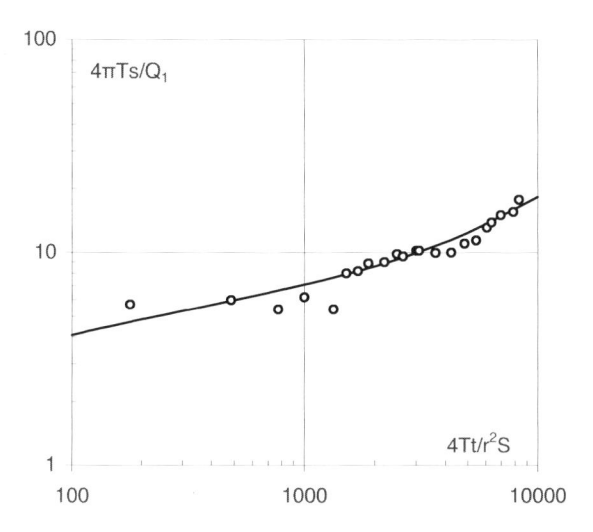

*Figure 11.19 The Beder well field. Type-curve match with drawdown data from a monitoring well.*

## 11.5 A discharge curve with a maximum value

Often, historical discharge data from a well field shows an increase followed by a decrease in pumping rate. The following discharge function often matches such data,

$$f(t) = Q_1 + Q_m t e^{-kt} = Q_1\left(1 + \alpha t e^{-kt}\right) \tag{11.15}$$

where $Q_1$ is the initial discharge rate, $Q_m$ is an intermediate value with dimension $L^3/T^2$ and $\alpha = Q_m/Q_1$ (dimension $T^{-1}$). In this case, $Q_1$ is also the final discharge rate. Figure 11.20 shows the discharge curve.

Substituting Equation 11.15 into Equation 11.2 gives

$$s = \frac{Q_1}{4\pi T}\int_u^\infty \left(1 + \frac{\alpha r^2 S}{4T}\left(\frac{1}{u} - \frac{1}{x}\right)\right)\exp\left(-\frac{kr^2 S}{4T}\left(\frac{1}{u} - \frac{1}{x}\right)\right)\frac{e^{-x}}{x}dx$$

FRACTURED AQUIFERS

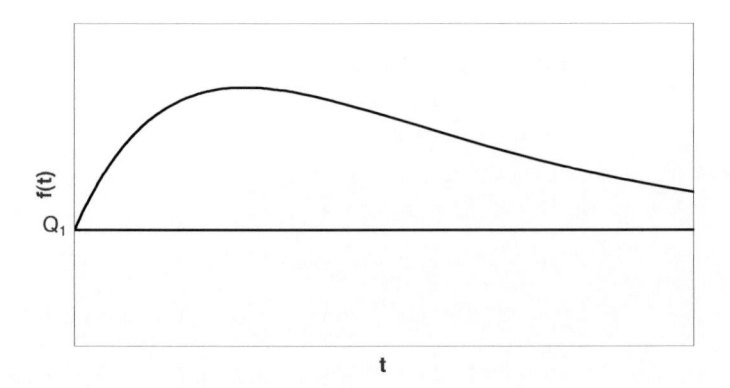

*Figure 11.20 Schematic of the discharge curve, Equation 11.15.*

After some manipulation, this equation reduces to

$$s = \frac{Q_1}{4\pi T}\left(W(u) + \frac{\alpha\beta^2}{4k}\exp\left(-\frac{\beta^2}{4u}\right)\left(\frac{W(u,\beta)}{4u} - \int_u^\infty \frac{\exp(-x-\beta^2/4x)}{x^2}dx\right)\right) \quad (11.16)$$

The time constant $k$ appears explicitly in Equation 11.16. Therefore, type curves are not general but dependent on the actual field data. Figure 11.21 shows type curves for $\alpha/k = 3.33$.

*Linderberg, (2006)* evaluated the integral on the right hand side of Equation 11.16 numerically.

For large values of $\beta$, i.e. large distances, the drawdown curves will be Theis curves.

Example 11.4

Data is from the Borreby well field operated by the municipal water supply of Odense, Denmark *(Nielsen, 2006)*. The wells tap a multilayer confined aquifer.

Figure 11.22 shows discharge data as a function of time. The discharge function chosen is

168

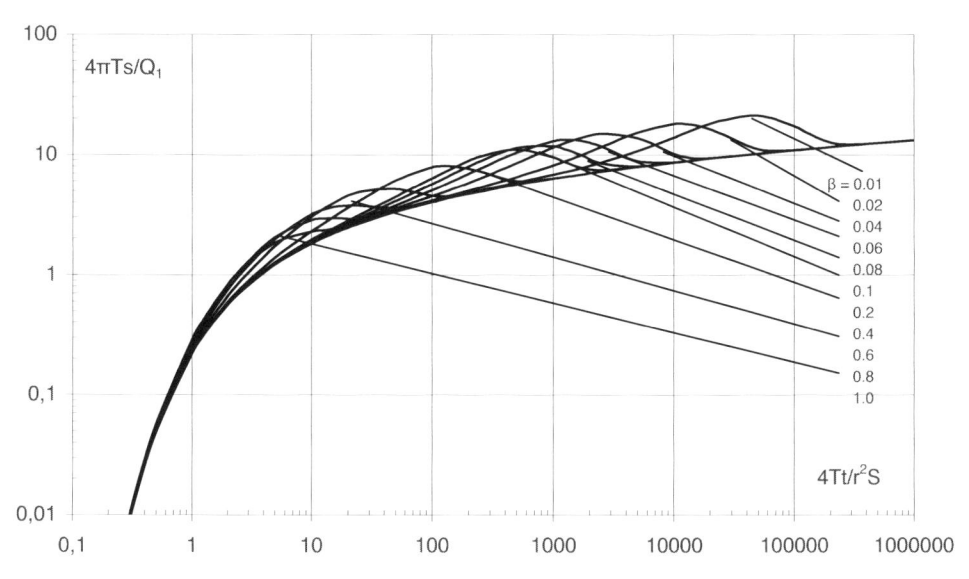

*Figure 11.21 Type curves, Equation 11.16, for $k/\alpha = 1/3$.*

$$f(t) = 90{,}000 + 9{,}000t \exp(-0.03t)$$

The type curves for this case are the ones shown in Figure 11.21. We have $\alpha = 9{,}000/90{,}000 = 0.1$ month⁻¹ and $k = 0.03$ month⁻¹

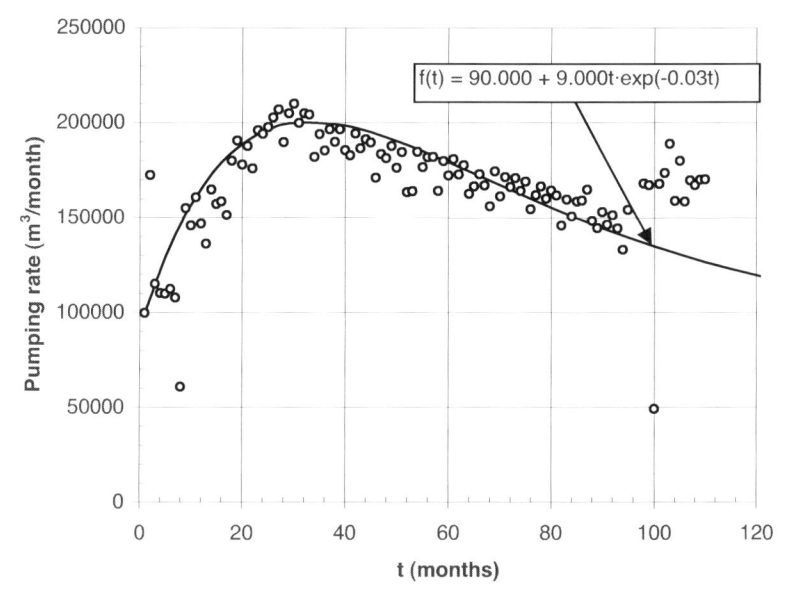

*Figure 11.22 Discharge data from the Borreby well field.*

Figure 11.22 indicates an acceptable data match up to about 96 months or 2900 days.

The table below shows drawdown data and Figure 11.23 shows a plot of drawdown data from a monitoring well at distance 240 m from the well field center.

| t (days) | s (m) | t (days) | s (m) | t (days) | s (m) | t (days) | s (m) |
|---|---|---|---|---|---|---|---|
| 39 | 2.2 | 796 | 6.301 | 1593 | 7.761 | 2348 | 7.601 |
| 65 | 2.761 | 828 | 6.651 | 1622 | 7.831 | 2376 | 7.681 |
| 71 | 2.681 | 857 | 6.731 | 1649 | 7.801 | 2409 | 7.661 |
| 92 | 2.43 | 893 | 6.921 | 1678 | 7.881 | 2441 | 7.62 |
| 128 | 2.86 | 920 | 7.091 | 1712 | 7.681 | 2472 | 7.821 |
| 163 | 3.061 | 948 | 7.161 | 1740 | 7.841 | 2496 | 7.601 |
| 183 | 3.161 | 988 | 7.131 | 1766 | 7.431 | 2529 | 7.671 |
| 248 | 4.741 | 1011 | 7.231 | 1815 | 7.7 | 2559 | 7.641 |
| 302 | 5.571 | 1037 | 7.151 | 1839 | 7.581 | 2588 | 7.731 |
| 337 | 5.261 | 1074 | 7.251 | 1860 | 7.601 | 2622 | 7.691 |
| 372 | 4.301 | 1101 | 7.501 | 1888 | 7.681 | 2650 | 7.62 |
| 397 | 4.501 | 1129 | 7.581 | 1921 | 7.7 | 2685 | 7.631 |
| 429 | 4.761 | 1170 | 7.651 | 1950 | 7.661 | 2711 | 7.761 |
| 458 | 4.831 | 1198 | 7.801 | 1983 | 7.711 | 2742 | 7.601 |
| 493 | 5.03 | 1226 | 7.561 | 2011 | 7.721 | 2781 | 7.631 |
| 520 | 5.161 | 1256 | 7.761 | 2041 | 7.681 | 2809 | 7.571 |
| 550 | 4.821 | 1282 | 7.931 | 2076 | 7.381 | 2844 | 7.241 |
| 582 | 5.731 | 1391 | 8.061 | 2103 | 7.711 | 2866 | 7.301 |
| 612 | 5.631 | 1403 | 8.2 | 2132 | 7.681 | 2895 | 7.231 |
| 646 | 5.971 | 1437 | 7.631 | 2173 | 7.391 | 2930 | 7.311 |
| 675 | 6.231 | 1466 | 8.04 | 2194 | 7.681 | 2962 | 7.151 |
| 702 | 6.231 | 1500 | 8.061 | 2258 | 7.53 | 2986 | 7.321 |
| 739 | 6.181 | 1523 | 7.911 | 2285 | 7.561 | | |
| 767 | 6.231 | 1555 | 7.901 | 2318 | 7.631 | | |

Matching with type curves, Figure 11.21, gives $s_{MP} = 0.54$ m, $t_{MP} = 0.67$ days and $\beta = 0.04$. The transmissivity is

$$T = \frac{Q_1}{4\pi s_{MP}} = \frac{90{,}000}{4\pi \cdot 0.54 \cdot 30{,}4 \cdot 24 \cdot 3600} = 0.00505 \text{ m}^2/\text{s}$$

The storage coefficient is

$$S = \frac{4Tt_{MP}}{r^2} = \frac{4 \cdot 0.00505 \cdot 0.67 \cdot 1440 \cdot 60}{240^2} = 0.0203$$

The long-term value of the storage coefficient is in the water table aquifer range. Figure 11.23 shows the type-curve match.

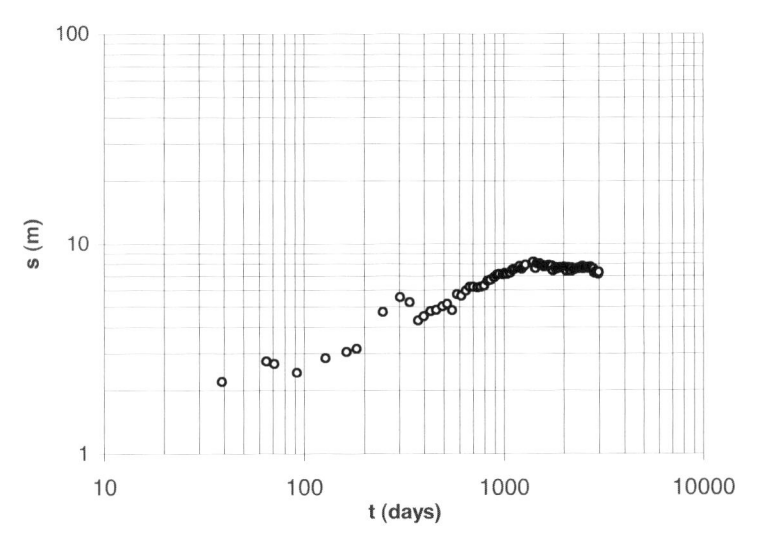

*Figure 11.23 The Borreby well field, drawdown data from a monitoring well.*

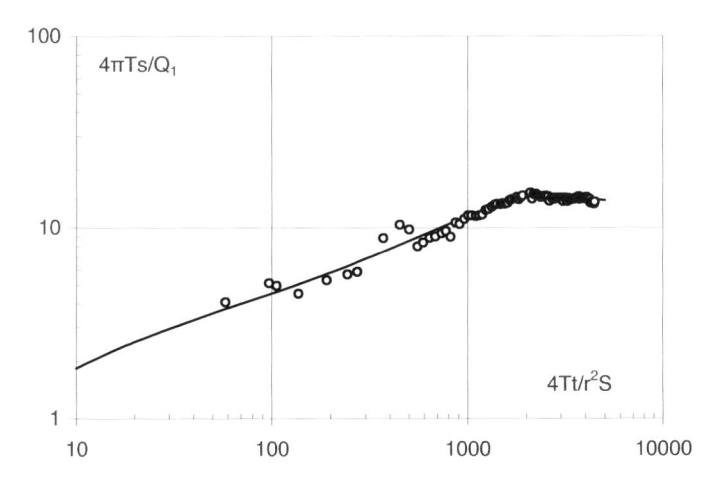

*Figure 11.24 The Borreby well field. Type-curve match with monitoring well data.*

# Literature

*Aguilera, R., 1980: Naturally Fractured Reservoirs. Penn Well Books.*

*Bourdet, D. and A. C. Gringarten, Sept. 1980: Determination of Fissured Volume and Block Size in Fractured Reservoirs by Type-curve Analysis. Paper SPE 9293 presented at the 1980 Annual Technical Conference and Exhibition, Dallas, p. 21-24.*

*Cinco-Ley, H. and V. F. Samaniego, September 1981: Transient Pressure Analysis for Fractured Wells. Journal of Petroleum Technology, p. 1749-1766.*

*Cinco, L. H., Samaniego, V. F. and A. N. Dominguez, August 1978: Transient Pressure Behavior for a Well with a Finite-conductivity Vertical Fracture. Society of Petroleum Engineers Journal, p. 253-264.*

*Cinco, L. H., Samaniego, V. F. and A. N. Dominguez, 1976: Unsteady-State Flow Behavior for a Well near a Natural Fracture. 51st Annual Fall Techn. Conf. and Exhibition, Society of Petroleum Engineers of AIME, New Orleans.*

*Cooper, H. H. and C. E. Jacob, 1946: A generalized graphical method for evaluating formation constants and summarizing well-field history. Transactions, American Geophysical Union 27 p. 526-34.*

*de Swaan, A., June, 1976: Analytic Solutions for Determining Naturally Fractured Reservoir Properties by Well Testing. Society of Petroleum Engineers Journal, p. 117-122.*

*Earlougher, R. C., Jr., 1977: Advances in Well Test Analyses. Monograph no. 5, Society of Petroleum Engineers, AIME.*

*Edelman, J. G., 1947: On the calculation of ground water flows. Ph.D. Thesis, Technical University of Delft (in Dutch).*

*Geological Survey of Denmark, 1971: Well tests at the Spangsbjerg Area, Esbjerg. Unpublished report prepared for the Municipal Water Supply of Esbjerg, Denmark (in Danish).*

*Geological Survey of Denmark, 1975: Nordvand (in Danish).*

*Geological Survey of Denmark, 1974: Consequences of strip mining at Roerdal, (in Danish).*

*Geological Survey of Denmark, 1976: The Susaa-Vendebaek Area. Report prepared for the Water Supply of Copenhagen, (in Danish).*

*Goode, D. J. and L. A. Senior, 1998: Review of Aquifer Test Results for the Lansdale Area, Montgomery County, Pennsylvania, 1980-95. USGS Open-File Report 98-294.*

*Gradshteyn, I.S. and I.W. Ryzhik, 1965: Table of Integrals, Series and Products. Academic Press.*

*Gringarten, A. C., 1982: Flow-Test Evaluation of Fractured Reservoirs. Geological Society of America, Special Paper 189, p. 237-263.*

*Gringarten, A. C. and P. A. Witherspoon, 1972: A Method of Analyzing Pump Test Data from Fractured Aquifers. Int. Soc. Rock Mechanics and Int. Assoc. Eng. Geol., Proc. Symposium Rock Mechanics, Stuttgart, Vol. 3-B, p. 1-9.*

*Gringarten, A. C. and H. J. Ramey, Jr., Aug. 1974: Unsteady-State Pressure Distribution Created by a Well With a Single horizontal Fracture, Partial Penetration, or Restricted Entry. Society of Petroleum Engineers Journal, Vol. 257, p. 413-426.*

*Gringarten, A. C. and H. J. Ramey, Jr. Oct. 1973: The Use of Point Source Solutions and Green's Functions for Solving Unsteady Flow Problems in Reservoirs. Society of Petroleum Engineers Journal, Vol. 225, p. 285-296.*

*Gringarten, A. C., Ramey, H. J. Jr. and R. Raghaven, August 1974: Unsteady-State Pressure Distributions Created by a well With a Single Infinite Conductivity Vertical Fracture. Society of Petroleum Engineers Journal, Vol. 14 No. 4, p. 347-360.*

*Gringarten, A. C., Ramey, H. J. Jr. and R. Raghaven, July 1975: Applied Pressure Analysis for Fractured Wells. Journal of Petroleum Technology (July), p.887-892.*

*Gringarten, A.C., 1984: Interpretation of Tests in Fissured and Multilayered Reservoirs With Double-Porosity Behavior: Theory and Practice. Journal of Petroleum Tecnology (April), p. 549-564.*

*Hantush, M.S., 1959: Analysis of Data From Pumping Tests In Leaky Aquifers. Transactions, American Geophysical Union, Vol. 37, No.6, p. 702-714.*

*Hantush, M.S., 1964a: Drawdown around Wells of Variable Discharge. Journal of Geophysical Research, Vol.69, No. 20.*

*Hantush, M.S., 1964b: Hydraulics of Wells. Advances in Hydroscience, Vol. 1.*

*Hawkins, M. F., 1956: A note on the skin effect. Trans. AIME, Vol. 207, p. 356-357.*
*Houlberg, R., M. Zimmermann, J. Hartman, A.G. Christensen, B. Blem and K.A. Nielsen, 1993: Analysis of historical monitoring data. Progress Report No. 70, Danish Environmental Protection Agency (in Danish).*

*Hurst, W., 1953: Establishment of the skin effect and its impediment to fluid flow in a well bore. Petroleum Engineering, V. 25, p. B-6 through B-16.*

*Jenkins, D. N. and J. K. Prentice, 1982: Theory for Aquifer Test Analysis in Fractured Rocks Under Linear (Nonradial) Flow Conditions. Ground Water Vol. 20, No. 1, p. 12-21.*

*Kazemi, H., December, 1969: Pressure Transient Analysis of Naturally Fractured Reservoirs with Uniform Fracture Distribution. Society of Petroleum Engineers Journal, p. 451-462.*

*Kazemi, H., Seth, M. S. and G. W. Thomas, December, 1969: The Interpretation of Interference Tests in Naturally Fractured Reservoirs with Uniform Fracture Distribution. Society of Petroleum Engineers Journal, p. 463-472.*

*Kruger A/S, 1998: Well tests at Aashoeje/Svansbjerg. Report prepared for the Municipal Water Supply of Koege, Denmark, (in Danish).*

*Kruger A/S, 2000: New well field at Gjeddesdal. Report prepared for the Water Supply of Greve Strand, Denmark (in Danish).*

*Kruseman, G. P. and N. A. de Ridder, 2001: Analysis and evaluation of pumping test data. ILRI Publication 47.*

*Linderberg, J., 2006: Numerical evaluation of an integral. University of Arhus, Denmark, Personal communication.*

*Louis, C.,1969: A Study of Ground water Flow in Jointed Rock and its influence on the Stability of Rock Masses. Imperial College Rock Mech. res. Report No. 10, London.*

*Moench, A. F., 1984: Double-porosity models for a Fissured Ground water Reservoir with Fracture Skin. Water Resources Research, Vol. 20 No 7, pp 831-846.*

*Najurieta, H. L., July, 1980: A theory for Pressure Transient Analysis in Naturally Fractured Reservoirs. Journal of Petroleum Technology, p. 1241-1250.*

Nielsen, K. A., 2007: *Analysis of drawdown data from Nybro. Report prepared for The Region of Roskilde, Denmark (in Danish).*

Nielsen, K. A., 2006 a: *Analysis of historical data from Borreby and Holmehave Well Fields. Report prepared for The Municipal Water Supply of Odense, Denmark (in Danish).*

Nielsen, K. A., 2005 a: *Analysis of drawdown data, Smaalyngen, Bornholm. Report prepared for The Region of Bornholm, Denmark, unpublished (in Danish).*

Nielsen, K. A., 2005 b: *Analysis of drawdown data, Albertslund. Report prepared for the County of Copenhagen, unpublished (in Danish).*

Nielsen, K. A., 2004 a: *Analysis of well test data at Snubbekorsgaard. Reports prepared for the Municipality of Hoeje Taastrup and the County of Copenhagen, unpublished, Denmark, (in Danish).*

Nielsen, K. A., 2004 b: *Analysis of drawdown data at Greve, Report prepared for the County of Roskilde, Denmark, unpublished (in Danish).*

Nielsen, K. A. 2006 c: *Well tests at Roedovre. Report prepared for the Municipal Water Supply of Rødovre, Denmark, unpublished (in Danish).*

Nielsen, K. A., 2006 b: *Analysis of drawdown data at Oesterparken. Report prepared for the Minicipality of Hoeje Taastrup, Denmark, unpublished, (in Danish).*

Raghaven, R., Uraiet A., and G. W. Thomas, August 1978: *Vertical Fracture Height: Effect on Transient Flow Behavior. Society of Petroleum Engineers Journal,265-277.*

Ramey, J. R., 1982: *Well-Loss Function and the Skin Effect: A Review In Recent Trends in Hydrogeology, edited by N.T Narasimhan, Geological Society of America, Boulder, Colorado.*

Rosenbeck, E., 1974: *Hydrogeology and hydrochemistry of the Ringsted-Haslev area. Master Thesis, University of Copenhagen, Denmark (in Danish).*

Russell, D. G. and N. E. Truitt, Oct. 1964: *Transient Pressure Behavior in Vertically Fractured Reservoirs. J. Pet. Tech., Vol. 231, p. 1159-1170.*

Scheidegger, A., 1960: *The Physics of Flow through Porous Media. University of Toronto Press.*

*Sen, Z. and A. Altunkaynak, October 2004: Variable Discharge Type-curve Solutions for Confined Aquifers. Journal of the American Water Resources Association, p. 1189-1196.*

*Serra, K., Reynolds, A. D. and R. Raghaven, December, 1983: New Pressure transient Analysis Methods for Naturally Fractured reservoirs. Journal of Petroleum Technology, p. 2271-2283.*

*Sharp, J. C., 1970: Fluid Flow Through Fissured Media. Ph.D. Thesis, University of London.*

*Sharp, J. C. and Y. N. T. Maini, 1972: Fundamental Considerations on the hydraulics of Joints in Rock. Intern. Soc. for Rock Mech. Symposium, Stuttgart.*

*Streltsova-Adams, T. D., 1978: Well hydraulics in heterogeneous aquifer formations. Advances in Hydroscience, Vent e Chow Ed.. Academic Press, New York.*

*Streltsova, T. D., 1976: Hydrodynamics of Ground water Flow in a Fractured Formation. Water Resources Research, Vol. 12 No. 3.*

*Sorensen, T., 1981: Permeability of Danish limestone in relation to geohydrology and water supply. The Institute of Engineering Geology, The Technical University of Denmark. In Danish.*

*Tam, V. T., F. De Smedt, O, Batellaan and A. Dassargues, 2003: Interpretation and analysis of aquifer tests in fractured-carstified carbonate limestone. Ground water in Fractured Rocks, Prague.*

*Theis, C. V., 1935: The lowering of the piezometer surface and the rate and discharge of a well using ground-water storage. Transactions, American Geophysical Union 16 p. 519-24.*

*Uldrich, D. O. and I. Ersaghi, October, 1979: A Method for Estimating the Interporosity flow parameter in Naturally Fractured Reservoirs. Society of Petroleum Engineers Journal, p. 324-332.*

*Utah Department of Natural Resources, 2000: Geohydrological and numerical simulation of ground-water flow in the central Virgin River basin of Iron and Washington Counties. Technical publication No. 116.*

*van Everdingen, A. F., 1953: The skin effect and its influence on the productive capacity of a well. American Institute of Mining Engineers Transactions, V. 198, p. 171-176.*

van Everdingen, A. F. and L. J. Meyer, April 1971: Analysis of Buildup Curves Obtained after Well Treatment. J. Pet. Tech., Vol. 251, p. 513-524.

Warren, J. E. and P. J. Root, September, 1963: The behavior of naturally fractured reservoirs. Society of Petroleum Engineers Journal, p. 245-255.

# List of symbols

| | |
|---|---|
| $e$ | Fracture aperture |
| $erf(z)$ | $= \dfrac{2}{\sqrt{\pi}} \displaystyle\int_0^z e^{-x^2} dx$, error function |
| $g$ | Acceleration of gravity |
| $h$ | Aquifer thickness |
| $h_f$ | Fracture width |
| $h_m$ | Thickness of matrix unit, or length of matrix block |
| $l$ | Characteristic length of matrix geometry, normally equal to $h_m$ |
| $n$ | Number of normal sets of fractures. $n=1$ for layers, and $n=3$ for blocks |
| $q$ | Flow in fracture |
| $r$ | Distance |
| $r_{ef}$ | Effective radius |
| $r_f$ | Radius of circular horizontal fracture |
| $r_i$ | Radius of influence for fracture |
| $r_s$ | Thickness of skin zone |
| $r_w$ | Nominal screen radius |
| $r_c$ | Casing radius |
| $r_{rm}$ | Rising main radius |
| $s$ | Drawdown in a monitoring well |
| $s_D$ | $= 4\pi Ts/Q$, dimensionless drawdown in a monitoring well |
| $s_{wD}$ | $= 4\pi Ts_w/Q$, dimensionless drawdown in the pumped well |
| $s_w$ | Drawdown in the pumped well |
| $s_w'$ | Residual drawdown |
| $s_w''$ | Recovery |
| $s_{MP}$ | Data curve Match Point co-ordinate, monitoring well |
| $s_{wMP}$ | Data curve Match Point co-ordinate, pumped well |
| $s_s$ | Skin loss |
| $\Delta s, \Delta s_w$ | Drawdown per log cycle |
| $t$ | Time since pumping started |
| $t_0$ | Intersection with zero drawdown axis |
| $t_D$ | $= Tt/r^2 S$ or $4Tt/r^2 S$, dimensionless time |

| | |
|---|---|
| $t_p$ | Time when pumping stops |
| $t_{MP}$ | Data curve Match Point co-ordinate |
| $\Delta t$ | Time since stop of pumping |
| $t_{tr}, \Delta t_{tr}$ | Time for end of transition period |
| $u$ | $= \dfrac{r^2 S}{4Tt}$ |
| $u_{CO}$ | $= u_{COL}\left(\dfrac{\varphi}{1+\varphi} + \dfrac{1}{1+\varphi}\sqrt{\dfrac{1}{\psi u_{COL}}}\tanh\sqrt{\psi u_{COL}}\right)$ |
| $u_{COL}$ | $= \dfrac{r^2(S_f + S_m)}{4T_f t}$, |
| $u^*$ | $= \dfrac{T_f t}{S_f + \xi S_m}$ |
| $w$ | Fracture width |
| $x_e$ | Aquifer extent |
| $x_f$ | Vertical fracture half-length |
| $B$ | Factor for formation loss, or |
| | $= \sqrt{\dfrac{T_f h_m}{2K_m}}$ for dual-porosity aquifer (Streltsova), or |
| | $= \dfrac{2}{\pi}\sqrt{\dfrac{T_f h_m^2}{T_m}}$ for dual-porosity aquifer, drawdown in matrix (Streltsova). |
| $C$ | Factor for second-order well loss |
| $C_i$ | Clogging index |
| $C_f$ | Clogging factor |
| $-E(-u)$ | Exponential integral (well function) |
| $I$ | Hydraulic gradient |
| $K$ | Hydraulic conductivity |
| $K_0(z)$ | Modified Bessel function of the second kind, zero order |
| $K_r$ | Horizontal hydraulic conductivity |
| $K_f$ | Fracture hydraulic conductivity |
| $K_z$ | Vertical hydraulic conductivity |
| $K_m$ | Matrix Hydraulic conductivity |
| $K_s$ | Skin zone hydraulic conductivity |
| $Q$ | Well yield |
| $R$ | Reynolds number |

| | |
|---|---|
| $S$ | Storage coefficient |
| $S_{CO}$ | $= \dfrac{T_f}{\eta_{CO}}$ |
| $S_s$ | Specific storage coefficient ($= S/h$) |
| $S_f$ | Fracture storage |
| $S_m$ | Matrix storage |
| $T$ | Transmissivity |
| $T_f$ | Fracture transmissivity |
| $T_m$ | Matrix transmissivity |
| $T_0$ | Pressure-dependent transmissivity prior to pumping |
| $W(u)$ | Well function $= \int_u^\infty \dfrac{e^{-x}}{x} dx$ |
| $\alpha$ | Constant related to pressure-dependent transmissivity, or $= \dfrac{4n(n+2)}{l^2} = \dfrac{4n(n+2)}{h_m^2}$, in a dual-porosity aquifer, or $= Q_2/Q_1$, Variable discharge case |
| $\beta$ | $= \dfrac{\alpha Q}{4\pi T_0}$, Pressure-dependent transmissivity, or $= kr^2 S/4T$, Variable discharge case |
| $\gamma$ | The exponential of Euler's constant ($\exp(0.5772) = 1{,}781$) |
| $\delta s_w$ | Vertical displacement between early and late semi-log straight line |
| $\varepsilon$ | $= \dfrac{3h_m^2 S_m T_f}{\gamma K_m h_m r^2 (3S_f + S_m)}$ |
| $\eta$ | $= T/S$, diffusivity |
| $\eta_m$ | $= T_m/S_m$, matrix diffusivity |
| $\eta_f$ | $= T_f/S_f$, fracture diffusivity |
| $\eta_{CO}$ | $= \dfrac{T_f}{S_f + S_m \sqrt{\dfrac{t}{\tau}} \tanh \sqrt{\dfrac{\tau}{t}}}$, Najurieta's diffusivity, stratum case, or $= \dfrac{T_f}{S_f + S_m \left(\sqrt{\dfrac{t}{\tau}} \coth \sqrt{\dfrac{\tau}{t}} - \dfrac{t}{\tau}\right)}$, Najurieta's diffusivity, blocks case |

| | | |
|---|---|---|
| $\eta_{COL}$ | $= \dfrac{T_f}{S_f + S_m}$, stratum case | |
| | $= \dfrac{3T_f}{3S_f + S_m}$, blocks case | |
| $\lambda$ | Inter-porosity flow factor $= \alpha r_w^2 \dfrac{K_m}{K_f}$ | |
| $\mu$ | Dynamic viscosity | |
| $\xi$ | Parameter describing matrix storage, (= 0 for early data, and = 1 for late data) | |
| $\rho$ | Density | |
| $\sigma$ | Skin factor | |
| $\sigma_p$ | Pseudo-skin factor | |
| $\tau$ | $= \dfrac{h_m^2}{4\gamma \eta_m}$ | |
| $\varphi$ | $= \dfrac{S_f}{S_m}$ | |
| $\psi$ | $= \dfrac{h_m^2 S_m T_f}{\gamma r^2 K_m h_m (S_f + S_m)}$ | |
| $\omega$ | $= \dfrac{S_f}{S_f + S_m}$ | |

# Index

**A**
Aalborg 22
Aarhus 164
Acceleration of gravity 13
Albertslund 193

**B**
Barrier boundary 40
Beder 164
Bessel function 49
Bilinear flow 102
Block length 52
Block size 82
Blocks case 78
Blocks model 35
Bornholm 20
Borreby

**C**
Confining layer 54
Casing radius 29
Clogging factor 22
Clogging index 21
Continuous line source 150

**D**
Damage factor 22
Damage ratio 21
Darcy's law 13
Density 13
Diagnostics 141
Diffusivity 42, 63
Dimensionless drawdown 10
Dimensionless time 10
Discharge rate 9
Dual-porosity aquifer 35
Dynamic viscosity 13

**E**
Effective radius 17
-Vertical fracture 96, 104
-Horizontal fracture 134
Error function 87
Esbjerg 59
Euler's constant 39

Exponential integral 9

**F**
Fluid velocity 13
Formation loss factor 21
Fracture aperture 13, 100
Fracture length 86
Fracture systems 14

**G**
Gjeddesdal 191, 208
Grapevine Pass aquifer 143
Greve 105

**H**
Half-fracture length 86
Hoeje Taastrup 27, 45, 198
Holmehave 157
Homogeneous aquifer 9
Horizontal fracture 129
Hydraulic fractureing 109
Hydraulic gradient 13

**I**
Ideal fracture 13
Inflection point 39
Inter-porosity flow factor 36

**J**
Jacob-Cooper equation 10

**K**
Koege 110

**L**
Layer model 35
Layer thickness 82
Leaky aquifer 41, 49
Linear flow
-Vertical fracture 102
-Horizontal fracture 130
Line source solution 9
Log cycle slope 19

## M
Match Point 50
Matrix layer thickness 52

## N
Natural fractures 14
Naturally fractured reservoir 35
Navier Stoke equations 13
Nybro 203

## O
Oerslev 195

## P
Partial penetration 23
-Vertical fracture 109
Phreatic aquifer 110
Plane sink 120
Pressure dependency 31
Pseudo-radial flow
-Vertical fracture 88
-Horizontal fracture 130
Pseudo-skin loss 23
Psedo-skin factor, horizontal fracture 134
Pseudo-steady state 37

## R
Radius of influence
-Vertical fracture 105
-Horizontal fracture 133
Recovery 25
Roedovre 102
Residual drawdown 25
Reynolds number 13
Rising main radius 29
Rough fracture 14

## S
Sampling frequency 147
Skin effect 17
Skin factor 17
-Vertical fracture 96, 104
-Horizontal fracture 135
Skin loss 17
Smooth fracture 14
Stimulated well 18
Storage coefficient 9
Stratum case 62

Strip aquifer 122

## T
Test duration 148
Theis solution 9
Transition time 69
Transmissivity 9
Turbulent flow 14, 17

## U
Under-stimulated well 18

## V
Variable discharge 149
-Exponential decrease 152
-Exponential increase 156
-Hyperbolic decrease 159
-Hyperbolic increase 162
-Linear increase 162
-Variation with a maximum value 167
Vertical dike 120
Vertical displacement 38
Vertical fracture 85
-Finite hydraulic conductivity 85, 100
-Infinite hydraulic conductivity 85, 94
-Uniform flux 85, 87
Vietnam 190

## W
Well bore storage 29
Well efficiency 21
Well loss factor 21
Well radius 9
Well function 9
Well performance 17

## Y
Yucca Mountain 188

# Annex 1. Problems

Problem 2.1 Type-curve and straight-line methods

Problem 4.1 Calculation of skin and well performance indicators

Problem 6.1 Warren and Root equations

Problem 6.2 Fractured aquifer, Yucca Mountain

Problem 6.3 Fractured aquifer, Vietnam

Problem 6.4 Fractured aquifer, Gjeddesdal

Problem 6.5 Fractured aquifer, Albertslund

Problem 6.6 Fractured aquifer, Oerslev

Problem 6.7 Fractured aquifer, Hoeje Taastrup

Problem 7.1 Single fracture

Problem 7.2 Single fracture

Problem 8.1 Single fracture

Problem 9.1 Combination of fracture systems, Gjeddesdal

Problem 11.1 Linear discharge variation

FRACTURED AQUIFERS

## Problem 2.1

Type-curve and straight-line methods

The drawdown data for this problem is synthetic data from a monitoring well. Determine $T$ and $S$ using the type-curve method and the straight-line method. The discharge rate is $Q = 100$ m³/hr and the distance to the monitoring well is $r = 100$ m.

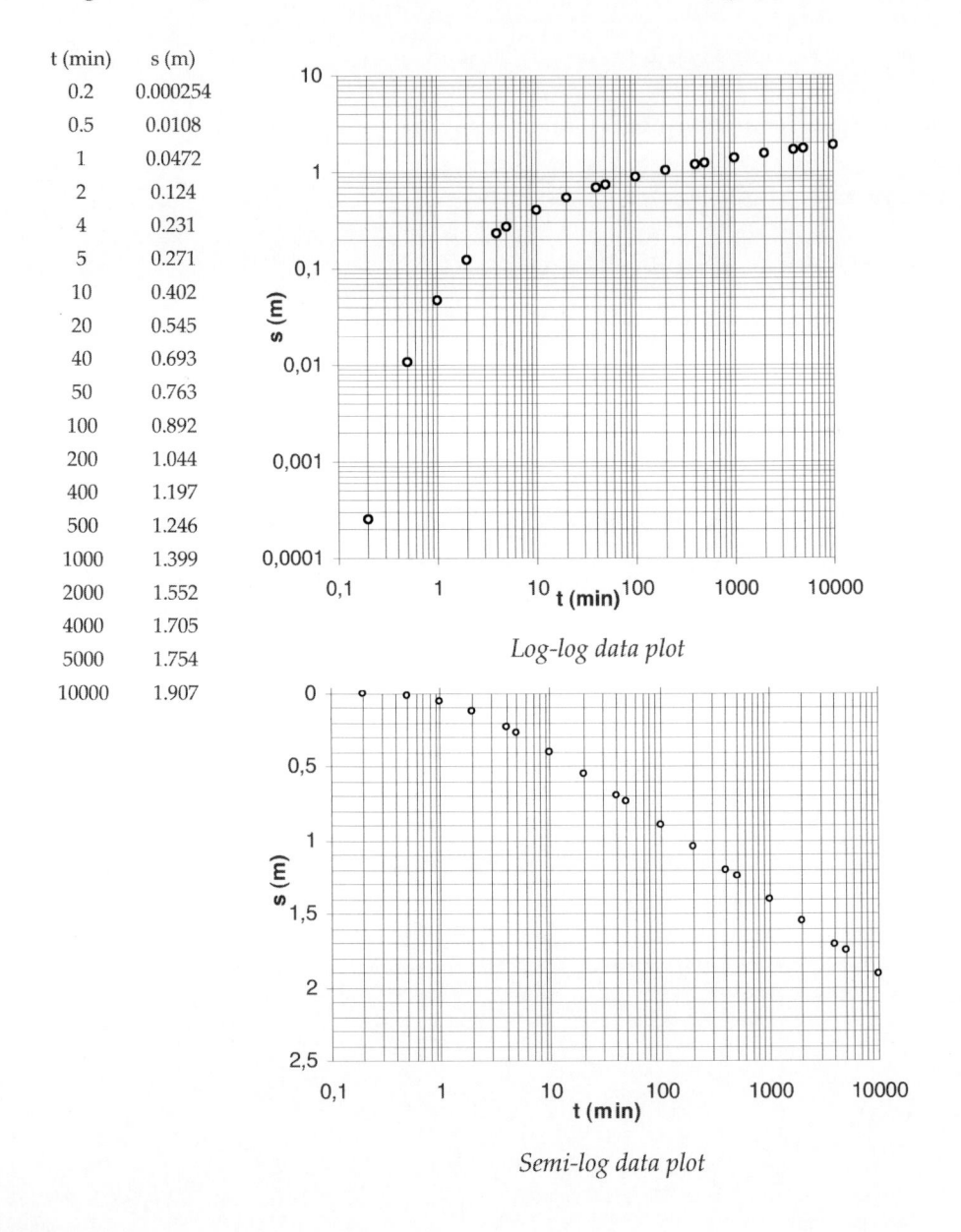

| t (min) | s (m) |
|---|---|
| 0.2 | 0.000254 |
| 0.5 | 0.0108 |
| 1 | 0.0472 |
| 2 | 0.124 |
| 4 | 0.231 |
| 5 | 0.271 |
| 10 | 0.402 |
| 20 | 0.545 |
| 40 | 0.693 |
| 50 | 0.763 |
| 100 | 0.892 |
| 200 | 1.044 |
| 400 | 1.197 |
| 500 | 1.246 |
| 1000 | 1.399 |
| 2000 | 1.552 |
| 4000 | 1.705 |
| 5000 | 1.754 |
| 10000 | 1.907 |

*Log-log data plot*

*Semi-log data plot*

## Problem 4.1

Calculation of skin and well performance indicators

The data given is synthetic. The table gives drawdown and recovery data from two different wells. For both wells, $Q = 72$ m³/hr and $r_w = 0.1$ m and $S = 0.001$. The drawdown before stop of pumping was $s_w = 8$ m and the duration of pumping was $t_p = 100.000$ min (recovery data).

Carry out the skin calculations and determine the well performance indicators.

| t (min) | sw (m) | (t_p +Δt)/Δt | sw −s''w (m) |
|---|---|---|---|
| 1 | 7.33 | 100001 | 3.05 |
| 2 | 7.606 | 50001 | 2.87 |
| 5 | 7.97 | 20001 | 2.62 |
| 10 | 8.245 | 10001 | 2.44 |
| 20 | 8.521 | 5001 | 2.26 |
| 50 | 8.885 | 2001 | 2.01 |
| 100 | 9.16 | 1001 | 1.83 |
| 200 | 9.436 | 501 | 1.65 |
| 500 | 9.8 | 201 | 1.40 |
| 1000 | 10.075 | 101 | 1.22 |
| 2000 | 10.351 | 51 | 1.04 |
| 5000 | 10.714 | 21 | 0.81 |
| 10000 | 10.99 | 11 | 0.64 |
| 20000 | 11.266 | 6 | 0.47 |

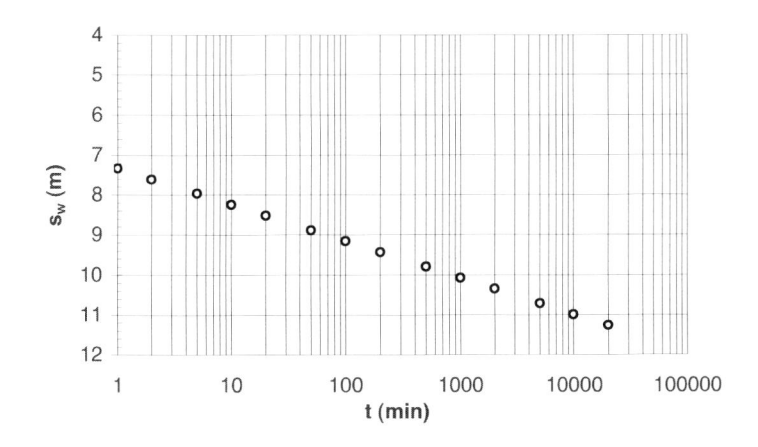

*Drawdown plot*

# FRACTURED AQUIFERS

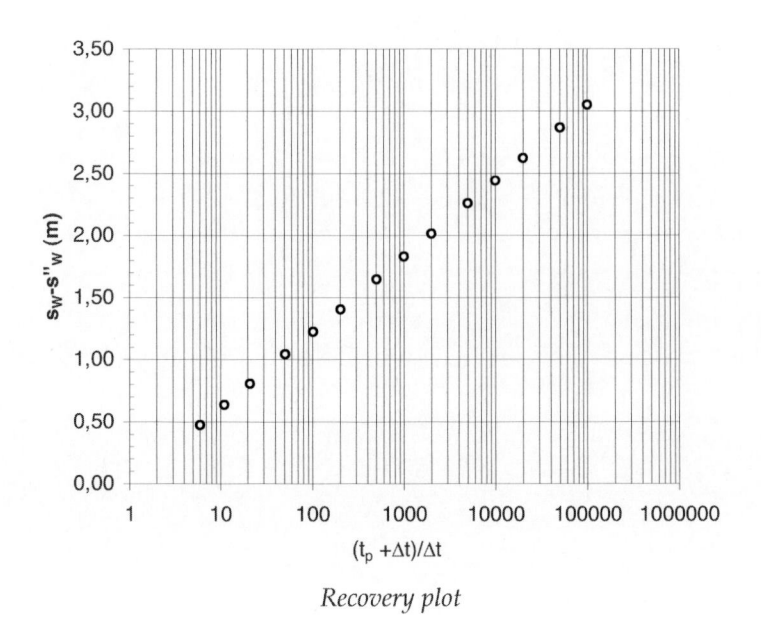

*Recovery plot*

**Problem 6.1**

Warren and Root equations

Show that Equations 6.4 and 6.9 may also be written, respectively

$$s_w = \frac{Q}{4\pi T_f}\left(2.3\log 2.25 t_D - W\left(\frac{\lambda t_D}{\omega(1-\omega)}\right) + W\left(\frac{\lambda t_D}{(1-\omega)}\right)\right)$$

and

$$s'_w = \frac{Q}{4\pi T_f}\left(2.3\log \frac{t_p + \Delta t}{\Delta t} + W\left(\frac{\lambda \Delta t_D}{\omega(1-\omega)}\right) - W\left(\frac{\lambda \Delta t_D}{(1-\omega)}\right)\right)$$

Next, show that Equations 6.7 and 6.8 are correct.

**Problem 6.2**

Fractured aquifer, Yucca Mountain

*Moench (1984)* published the drawdown data below from a well test in Yucca Mountain, USA. The well taps porous and fractured volcanic deposits. The well yield is $Q = 0.0358$ m³/s and the well radius is $r_w = 0.11$ m. The matrix layer thickness is $h_m = 1$ m as an estimate.

Prepare data diagnostics and calculate hydraulic parameters, skin and well performance indicators.

| t (min) | sw (m) | t (min) | sw (m) | t (min) | sw (m) |
|---|---|---|---|---|---|
| 0.05 | 2.1513 | 5 | 8.24 | 180 | 9 |
| 1 | 3.769 | 6 | 8.32 | 200 | 9.02 |
| 0.15 | 4.583 | 7 | 8.41 | 240 | 9.04 |
| 0.2 | 4.858 | 8 | 8.46 | 300 | 9.07 |
| 0.25 | 5.003 | 9 | 8.54 | 400 | 9.11 |
| 0.3 | 5.119 | 10 | 8.62 | 500 | 9.14 |
| 0.35 | 5.23 | 12 | 8.67 | 600 | 9.17 |
| 0.4 | 5.39 | 14 | 8.7 | 700 | 9.18 |
| 0.45 | 5.542 | 16 | 8.74 | 800 | 9.21 |
| 0.5 | 5.69 | 18 | 8.76 | 900 | 9.25 |
| 0.6 | 5.96 | 20 | 8.77 | 1000 | 9.3 |
| 0.7 | 6.19 | 25 | 8.81 | 1200 | 9.44 |
| 0.8 | 6.42 | 30 | 8.84 | 1400 | 9.55 |
| 0.9 | 6.59 | 35 | 8.84 | 1600 | 9.64 |
| 1 | 6.74 | 40 | 8.86 | 1800 | 9.74 |
| 1.2 | 6.96 | 50 | 8.86 | 2000 | 9.78 |
| 1.4 | 7.17 | 60 | 8.9 | 2200 | 9.8 |
| 1.6 | 7.33 | 70 | 8.91 | 2400 | 9.84 |
| 1.8 | 7.45 | 80 | 8.92 | 2600 | 9.93 |
| 2 | 7.56 | 90 | 8.93 | 2800 | 10.03 |
| 2.5 | 7.76 | 100 | 8.95 | 3000 | 10.08 |
| 3 | 7.93 | 120 | 8.97 | 3500 | 10.26 |
| 3.5 | 8.03 | 140 | 8.98 | 4000 | 10.3 |
| 4 | 8.12 | 160 | 8.99 | 4200 | 10.41 |

FRACTURED AQUIFERS

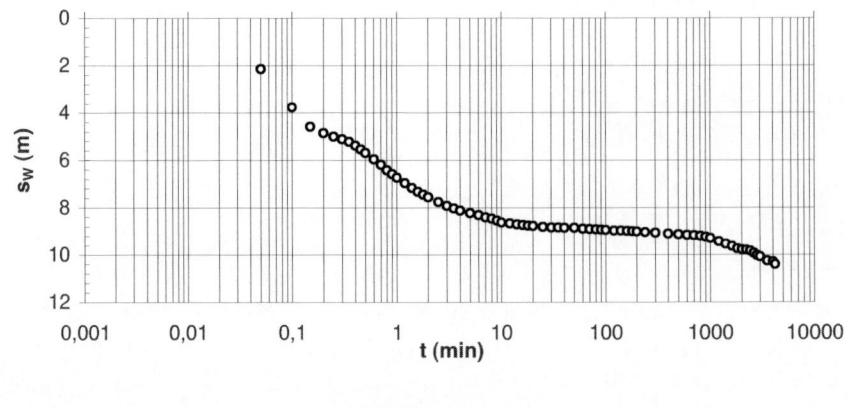

*Drawdown plot*

**Problem 6.3**

Fractured aquifer, Vietnam

*Tam et al (2003)* published recovery data after a well test in Vietnam. The well taps fractured and karstic limestone. The well yield was $Q = 12.7$ m³/hr and $r_w = 0.1125$ m. $S_f + S_m = 0.1$ (estimate). The duration of the test was $t_p = 1320$ min. The drawdown at the end of the pumping test was $s_w = 9.94$ m. Prepare data diagnostics and calculate hydraulic parameters, skin and well performance indicators.

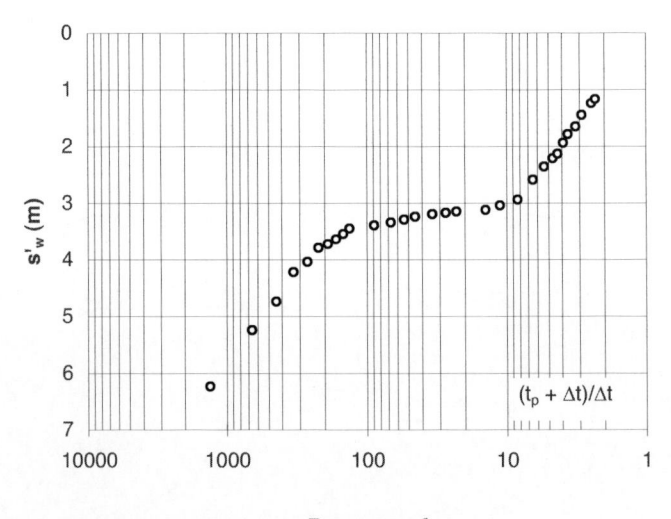

*Recovery plot*

| $(t_p+\Delta t)/\Delta t$ | $s'_w$ (m) | $(t_p+\Delta t)/\Delta t$ | $s'_w$ (m) |
|---|---|---|---|
| 1321 | 6.24 | 27.4 | 3.17 |
| 661 | 5.24 | 23 | 3.15 |
| 441 | 4.74 | 14.2 | 3.12 |
| 331 | 4.22 | 11.15 | 3.04 |
| 265 | 4.04 | 8.33 | 2.94 |
| 221 | 3.79 | 6.5 | 2.59 |
| 189 | 3.72 | 5.4 | 2.36 |
| 166 | 3.64 | 4.67 | 2.21 |
| 147.67 | 3.54 | 4.3 | 2.13 |
| 133 | 3.445 | 3.93 | 1.94 |
| 89 | 3.39 | 3.64 | 1.79 |
| 67 | 3.34 | 3.2 | 1.65 |
| 53.8 | 3.29 | 2.89 | 1.45 |
| 45 | 3.24 | 2.47 | 1.24 |
| 34 | 3.19 | 2.32 | 1.17 |

## Problem 6.4

Fractured aquifer, Gjeddesdal

The table below shows unpublished data from a drawdown test. The well taps a fractured limestone aquifer at Gjeddesdal south of Copenhagen, Denmark *(Kruger, 2000)*. The well yield was $Q = 90.5$ m³/hr and the well radius $r_w = 0.1125$ m. The table includes drawdown data from 2 monitoring wells, 44 and 2691 at distances 490 m and 670 m respectively.

Carry out data diagnostics and calculate hydraulic parameters, skin and well efficiency for the pumped well. Determine aquifer hydraulic parameters from monitoring well data. Assume $h_m = 1$ m.

# FRACTURED AQUIFERS

|  | Pumped well |  |  |  | Well No. 44 | Well no. 2691 |
|---|---|---|---|---|---|---|
| t (min) | $s_w$(m) | t (min) | $s_w$(m) | t (min) | s (m) | s (m) |
| 0.2 | 0.445 | 90 | 0.943 | 30 | 0.044 | 0.06 |
| 0.3 | 0.452 | 120 | 0.973 | 40 | 0.06 | 0.08 |
| 0.4 | 0.471 | 150 | 0.98 | 50 | 0.068 | 0.092 |
| 0.48 | 0.49 | 180 | 1 | 60 | 0.08 | 0.11 |
| 0.58 | 0.5 | 210 | 1.014 | 80 | 0.095 | 0.12 |
| 0.75 | 0.528 | 240 | 1.018 | 120 | 0.105 | 0.15 |
| 0.9 | 0.547 | 300 | 1.037 | 180 | 0.13 | 0.17 |
| 1.1 | 0.566 | 400 | 1.074 | 230 | 0.145 | 0.18 |
| 1.3 | 0.577 | 600 | 1.12 | 300 | 0.16 | 0.198 |
| 1.5 | 0.596 | 800 | 1.15 | 360 | 0.17 | 0.2 |
| 1.8 | 0.603 | 1000 | 1.18 | 420 | 0.18 | 0.21 |
| 2 | 0.618 | 1200 | 1.22 | 600 | 0.21 | 0.23 |
| 2.5 | 0.641 | 1500 | 1.244 | 800 | 0.235 | 0.26 |
| 3 | 0.66 | 1800 | 1.263 | 1000 | 0.27 | 0.28 |
| 3.5 | 0.679 | 2000 | 1.289 | 1300 | 0.33 | 0.3 |
| 4 | 0.698 | 2500 | 1.338 | 1500 | 0.34 | 0.33 |
| 5 | 0.716 | 3000 | 1.4 | 2000 | 0.38 | 0.36 |
| 7 | 0.746 | 4000 | 1.47 | 2500 | 0.44 | 0.38 |
| 10 | 0.773 | 6000 | 1.53 | 3000 | 0.5 | 0.42 |
| 12 | 0.8 | 8000 | 1.6 | 4000 | 0.52 | 0.48 |
| 16 | 0.814 | 10000 | 1.697 | 6000 | 0.54 | 0.51 |
| 18 | 0.814 | 12000 | 1.697 | 8000 | 0.56 | 0.51 |
| 20 | 0.829 | 15000 | 1.8 | 10000 | 0.65 | 0.62 |
| 25 | 0.848 | 18000 | 1.847 | 12000 | 0.7 | 0.7 |
| 30 | 0.86 | 20000 | 1.847 | 15000 | 0.85 | 0.69 |
| 35 | 0.863 | 25000 | 1.922 | 18000 | 0.9 | 0.76 |
| 40 | 0.871 | 30000 | 2 | 20000 | 0.9 | 0.78 |
| 50 | 0.889 | 35000 | 2.073 | 27000 | 0.9 | 0.78 |
| 60 | 0.909 | 40000 | 2.093 | 30000 |  | 0.85 |
| 75 | 0.939 | 48000 | 2.15 | 37000 |  | 0.9 |

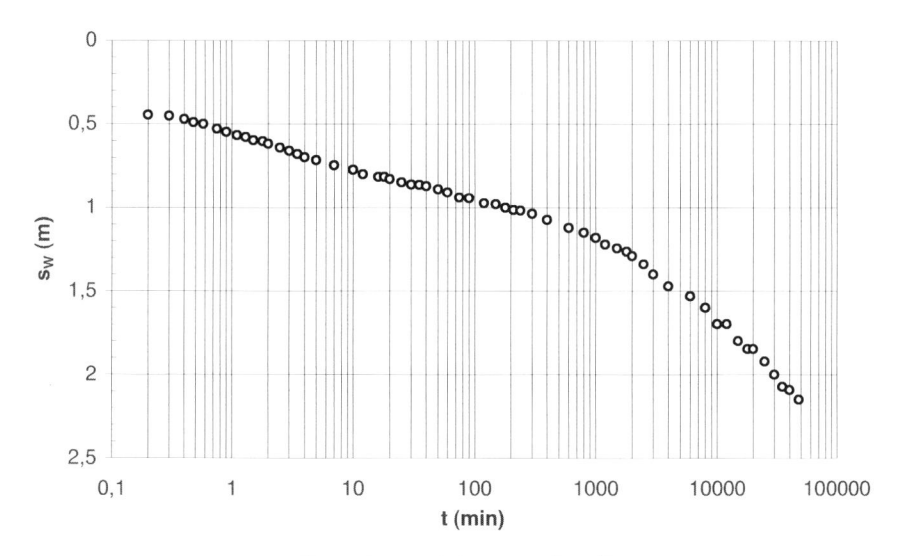

*Drawdown plot, pumped well*

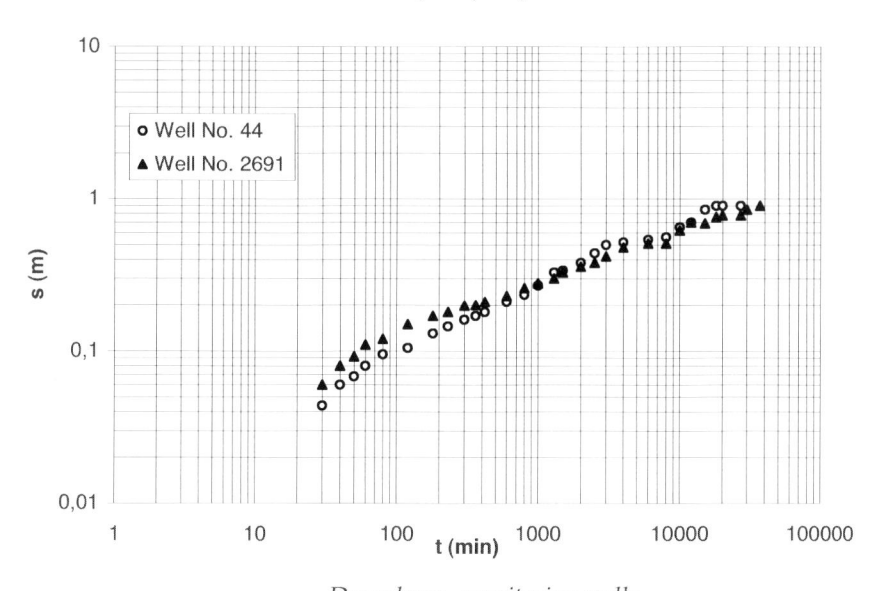

*Drawdown, monitoring wells*

## Problem 6.5

Fractured aquifer, Albertslund

The table below shows recovery from two monitoring wells data after a pumping test on a well field near Albertslund, Denmark *(Nielsen, 2005 b)*. The yield was 83.2 m³/hr. The distances from the well field center to monitoring wells 6 and 8 are 200 and 297 m. respectively. Changes in atmospheric pressure influence data. Find the best type-curve match and calculate aquifer hydraulic parameters.

| t (hours) | Well No. 6 s-s' (m) | Well No. 8 s-s' (m) | t (hours) | Well No. 6 s-s' (m) | Well No. 8 s-s' (m) |
|---|---|---|---|---|---|
| 1 | 0.015 | 0.005 | 36 | 0.149 | 0.115 |
| 2 | 0.034 | 0.021 | 37 | 0.152 | 0.11 |
| 3 | 0.042 | 0.028 | 38 | 0.157 | 0.119 |
| 4 | 0.059 | 0.039 | 39 | 0.162 | 0.112 |
| 5 | 0.061 | 0.04 | 40 | 0.151 | 0.116 |
| 6 | 0.078 | 0.048 | 41 | 0.169 | 0.125 |
| 7 | 0.07 | 0.048 | 42 | 0.157 | 0.117 |
| 8 | 0.097 | 0.055 | 43 | 0.163 | 0.123 |
| 9 | 0.1 | 0.056 | 44 | 0.155 | 0.118 |
| 10 | 0.11 | 0.062 | 45 | 0.147 | 0.117 |
| 11 | 0.101 | 0.062 | 46 | 0.15 | 0.122 |
| 12 | 0.11 | 0.065 | 47 | 0.142 | 0.108 |
| 13 | 0.114 | 0.083 | 48 | 0.157 | 0.114 |
| 14 | 0.111 | 0.073 | 49 | 0.147 | 0.123 |
| 15 | 0.121 | 0.079 | 50 | 0.16 | 0.128 |
| 16 | 0.122 | 0.086 | 51 | 0.155 | 0.124 |
| 17 | 0.127 | 0.087 | 52 | 0.151 | 0.122 |
| 18 | 0.128 | 0086 | 53 | 0.152 | 0.134 |
| 19 | 0.129 | 0.087 | 54 | 0.153 | 0.127 |
| 20 | 0.133 | 0.09 | 55 | 0.162 | 0.127 |
| 21 | 0.137 | 0.092 | 56 | 0.163 | 0.128 |
| 22 | 0.133 | 0.091 | 57 | 0.16 | 0.128 |
| 23 | 0.121 | 0.088 | 58 | 0.162 | 0.134 |
| 24 | 0.13 | 0.096 | 59 | 0.167 | 0.137 |
| 25 | 0.135 | 0.095 | 60 | 0.163 | 0.139 |
| 26 | 0.128 | 0.09 | 61 | 0.164 | 0.138 |
| 27 | 0.136 | 0.093 | 62 | 0.173 | 0.135 |
| 28 | 0.14 | 0.106 | 63 | 0.166 | 0.146 |
| 29 | 0.141 | 0.105 | 64 | 0.167 | 0.14 |
| 30 | 0.144 | 0.103 | 65 | 0.177 | 0.144 |
| 31 | 0.145 | 0.107 | 66 | 0.171 | 0.14 |
| 32 | 0.154 | 0.108 | 67 | 0.172 | 0.141 |
| 33 | 0.154 | 0.106 | 68 | 0.171 | 0.139 |
| 34 | 0.149 | 0.106 | 69 | 0.177 | 0.137 |
| 35 | 0.148 | 0.108 | 70 | 0.178 | 0.144 |
|  |  |  | 71 | 0.176 | 0.138 |

*Recovery plot*

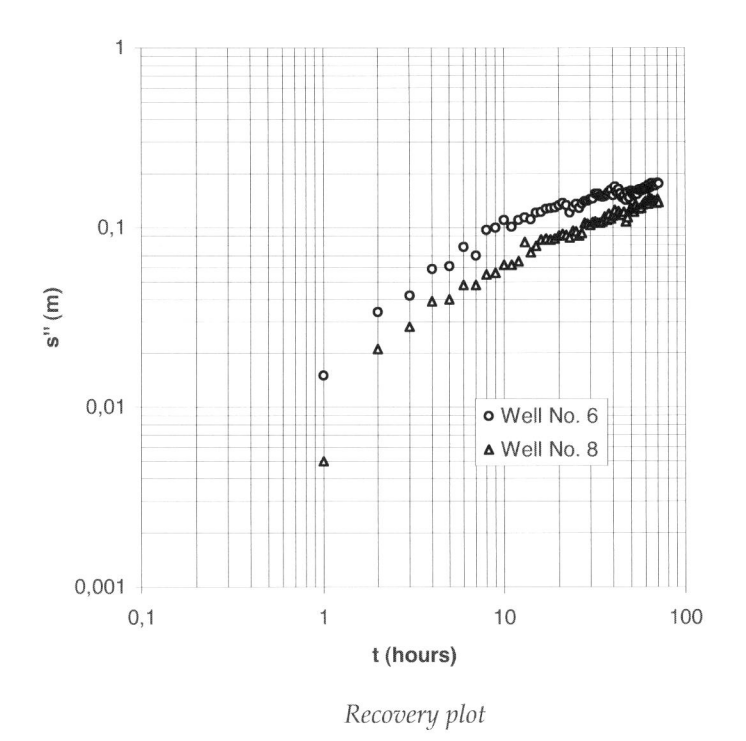

**Problem 6.6**

Fractured aquifer, Oerslev

The table shows data from a drawdown test on Well No. O1 situated at Oerslev in central Zealand, Denmark, *(Geological Survey of Denmark, 1976)*. The well yield is $Q = 73.5$ m³/hr end the well radius $r_w = 0.114$ m. The table includes data from three monitoring wells O5, 1336 and A11 at distances 920 m, 2350 m and 4740 m respectively. The aquifer is of Selandian age and consists of alternating layers of sand, clay and limestone.

Carry out data diagnostics, calculate hydraulic parameters and skin for the pumped well. Write the drawdown equation for the pumped well. Calculate aquifer hydraulic parameters using data from Well No. O5 and give a qualitative interpretation of data from the two other monitoring wells. Note, that time for monitoring wells is expressed as $t/r^2$. Hint: Use the same model for the pumped well and monitoring wells. Start with Well No. O5.

|  |  |  |  | Monitoring wells |  | O5 |  | 1336 |  | A11 |
| --- | --- | --- | --- | --- | --- | --- | --- | --- | --- | --- |
|  | Pumped well |  |  |  |  |  | t/r² |  | t/r² |  |
| t (min) | Sw (m) | t (min) | Sw (m) | t/r² (min/m²) | s (m) | (min/m²) | s (m) | (min/m²) | s (m) |
| 0.24 | 0.82 | 90 | 1.0625 | 0.0000043 | 0.0015 | 0.00002 | 0.002 | 0.000016 | 0.005 |
| 0.61 | 0.84 | 120 | 1.085 | 0.0000075 | 0.005 | 0.000027 | 0.003 | 0.000022 | 0.006 |
| 0.88 | 0.85 | 150 | 1.105 | 0.0000105 | 0.008 | 0.0000325 | 0.003 | 0.00003 | 0.008 |
| 1.2 | 0855 | 180 | 1.123 | 0.0000135 | 0.013 | 0.000038 | 0.0035 | 0.000036 | 0.015 |
| 1.5 | 0.86 | 240 | 1.155 | 0.0000185 | 0.02 | 0.000042 | 0.004 | 0.000042 | 0.016 |
| 2 | 0.87 | 300 | 1.17 | 0.000023 | 0.029 | 0.000049 | 0.004 | 0.000047 | 0.011 |
| 2.5 | 0.875 | 380 | 1.195 | 0.00003 | 0.037 | 0.000064 | 0.005 | 0.000053 | 0.011 |
| 3 | 0.88 | 420 | 1.205 | 0.0000355 | 0.0422 | 0.000077 | 0.005 | 0.000078 | 0.012 |
| 3.5 | 0.8825 | 540 | 1.225 | 0.000044 | 0.05 | 0.000087 | 0.01 | 0.000092 | 0.018 |
| 4 | 0.885 | 600 | 1.24 | 0.000053 | 0.06 | 0.00011 | 0.0115 | 0.00013 | 0.027 |
| 4.5 | 0.89 | 720 | 1.2625 | 0.000062 | 0.067 | 0.00012 | 0.013 | 0.00015 | 0.04 |
| 5 | 0.895 | 840 | 1.275 | 0000071 | 0.073 | 0.00013 | 0.014 | 0.00019 | 0.043 |
| 5.5 | 0.8975 | 960 | 1.285 | 0.00009 | 0.083 | 0.00016 | 0.016 | 0.000205 | 0.05 |
| 6 | 0.9 | 1080 | 1.2975 | 0.000112 | 0.09 | 0.00018 | 0.0175 | 0.00025 | 0.047 |
| 6.5 | 0.89865 | 1320 | 1.3125 | 0.000122 | 0.1 | 0.00019 | 0.0175 | 0.00028 | 0.049 |
| 7 | 0.9 | 1500 | 1.345 | 0.000148 | 0.11 | 0.00026 | 0.02 | 0.00032 | 0.054 |
| 7.5 | 0.9075 | 1800 | 1.355 | 0.00018 | 0.128 | 0.000295 | 0.027 | 0.00037 | 0.059 |
| 8 | 0.9125 | 2100 | 1.375 | 0.00021 | 0.132 | 0.00033 | 0.028 | 0.0004 | 0.069 |
| 8.5 | 0.917 | 2300 | 1.385 | 0.00025 | 0.14 | 0.00035 | 0.032 | 0.00043 | 0.078 |
| 9 | 0.92 | 2580 | 1.395 | 0.0003 | 0.155 | 0.00039 | 0.035 |  |  |
| 9.5 | 0.9225 | 2980 | 142 | 0.00033 | 0.16 | 0.00042 | 0.042 |  |  |
| 10 | 0.915 | 3300 | 1.435 | 0.00042 | 0.175 | 0.00049 | 0.043 |  |  |
| 13 | 0.93 | 4000 | 1.45 | 0.000495 | 0.195 | 0.00051 | 0.05 |  |  |
| 14 | 0.935 | 4300 | 1.455 | 0.00058 | 0.205 | 0.00059 | 0.047 |  |  |
| 16 | 0.9475 | 5000 | 1.453 | 0.00063 | 0.218 | 0.00066 | 0.048 |  |  |
| 18 | 0.945 | 5300 | 1.455 | 0.00071 | 0.225 | 0.0007 | 0.0595 |  |  |
| 20 | 0.954 | 6050 | 1.463 | 0.00078 | 0.23 | 0.00078 | 0.07 |  |  |
| 25 | 0.975 | 6600 | 1.481 | 0.00086 | 0.24 | 0.0008 | 0.075 |  |  |
| 30 | 0.9825 | 7500 | 1.493 | 0.001 | 0.25 | 0.00098 | 0.077 |  |  |
| 35 | 0.99 | 8100 | 1.5 | 0.0012 | 0.255 | 0.0013 | 0.081 |  |  |
| 40 | 1.004 | 8800 | 1.495 | 0.0014 | 0.28 | 0.0016 | 0.102 |  |  |
| 45 | 1.01 | 9800 | 1.52 | 0.002 | 0.29 | 0.00185 | 0.128 |  |  |
| 50 | 1.0225 | 10800 | 1.575 | 0.0023 | 0.31 |  |  |  |  |
| 60 | 1.0295 | 11400 | 1.585 | 0.0028 | 0,32 |  |  |  |  |
| 75 | 1.045 |  |  | 0.0033 | 0.33 |  |  |  |  |
|  |  |  |  | 0.0043 | 0.345 |  |  |  |  |
|  |  |  |  | 0.0052 | 0.35 |  |  |  |  |
|  |  |  |  | 0.0065 | 0.38 |  |  |  |  |
|  |  |  |  | 0.0088 | 0.385 |  |  |  |  |
|  |  |  |  | 0.012 | 0.4 |  |  |  |  |

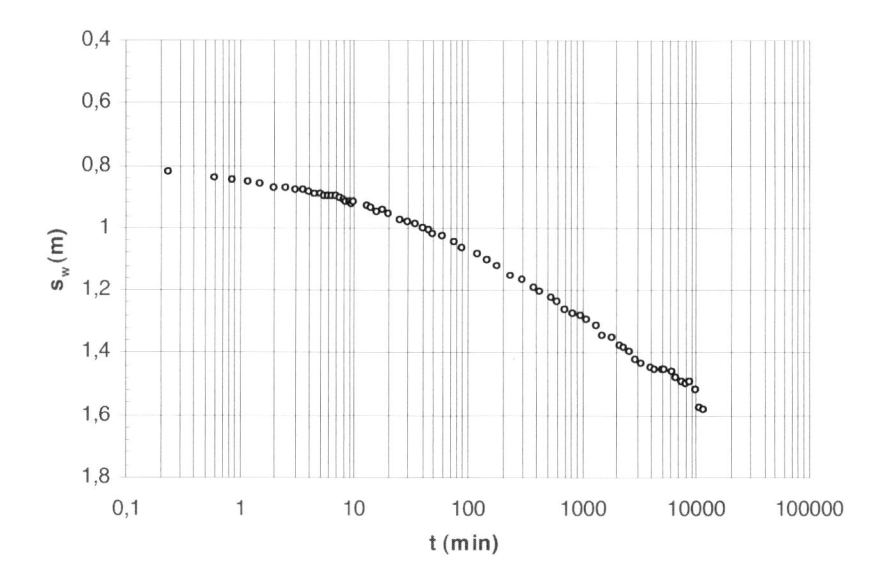

*Data plot, pumped well*

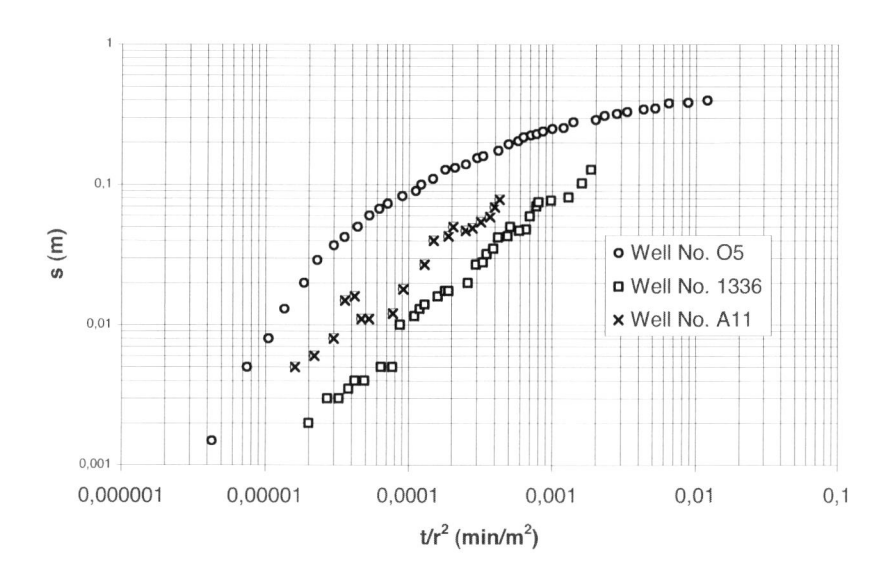

*Data plot, monitoring wells*

## Problem 6.7

Fractured Aquifer, Hoeje Taastrup

The table shows recovery data from the test evaluated in Examples 6.1 and 6.3, *(Nielsen, 2006 b)*. $Q = 15.1$ m³/h, $r_w = 0.1$ m. The distances to monitoring wells 16 and 17 are 47.5 and 32 m respectively. The duration of the drawdown test was $t_p = 354$ min, and the drawdown at the end of the test was $s_w = 0.1826$ m.

### The pumped well

| Δt (min) | $(t_p+\Delta t)/\Delta t$ | sw - sw' (m) | Δt (min) | $(t_p+\Delta t)/\Delta t$ | sw - sw' (m) | Δt (min) | $(t_p+\Delta t)/\Delta t$ | sw - sw' (m) |
|---|---|---|---|---|---|---|---|---|
| 0.03 | 10627.1 | 0.027 | 3.53 | 101.3 | 0.114 | 40.15 | 9.823 | 0.126 |
| 0.13 | 2657.7 | 0.057 | 4.03 | 88.83 | 0.114 | 45.15 | 8.846 | 0.126 |
| 0.23 | 1519.1 | 0.090 | 4.53 | 79.14 | 0.114 | 50.15 | 8.063 | 0.127 |
| 0.33 | 1063.7 | 0.097 | 5.03 | 71.38 | 0.114 | 55.15 | 7.423 | 0.130 |
| 0.43 | 818.5 | 0.100 | 5.53 | 65.02 | 0.114 | 60.15 | 6.889 | 0.134 |
| 0.53 | 665.2 | 0.104 | 6.03 | 59.71 | 0.117 | 70.15 | 6.050 | 0.138 |
| 0.63 | 560.3 | 0.104 | 7.03 | 51.36 | 0.117 | 80.15 | 5.420 | 0.138 |
| 0.73 | 484.0 | 0.107 | 8.03 | 45.1 | 0.121 | 90.15 | 4.929 | 0.139 |
| 0.83 | 426.1 | 0.110 | 9.03 | 40.21 | 0.121 | 100.15 | 4.537 | 0.136 |
| 0.93 | 380.5 | 0.114 | 10.03 | 36.31 | 0.124 | 120.15 | 3.948 | 0.137 |
| 1.03 | 343.8 | 0.114 | 12.03 | 30.44 | 0.124 | 140.15 | 3.528 | 0.139 |
| 1.23 | 288.2 | 0.114 | 14.03 | 26.24 | 0.124 | 160.15 | 3.212 | 0.143 |
| 1.43 | 248.1 | 0.114 | 16.03 | 23.09 | 0.125 | 180.15 | 2.966 | 0.144 |
| 1.63 | 217.9 | 0.114 | 18.03 | 20.64 | 0.125 | 200.15 | 2.770 | 0.145 |
| 1.83 | 194.2 | 0.114 | 20.03 | 18.68 | 0.125 | 220.15 | 2.609 | 0.147 |
| 2.03 | 175.2 | 0.114 | 22.03 | 17.08 | 0.125 | 240.15 | 2.475 | 0.148 |
| 2.23 | 159.6 | 0.114 | 24.15 | 15.67 | 0.125 | 260.15 | 2.362 | 0.149 |
| 2.43 | 146.6 | 0.114 | 26.15 | 14.55 | 0.125 | 280.15 | 2.264 | 0.150 |
| 2.63 | 135.5 | 0.114 | 28.15 | 13.58 | 0.125 | 300.15 | 2.180 | 0.151 |
| 2.83 | 126.0 | 0.114 | 30.15 | 12.75 | 0.125 | 350.15 | 2.012 | 0.154 |
| 3.03 | 117.8 | 0.114 | 35.15 | 11.08 | 0.126 | | | |

## Monitoring wells

| Δt (min) | Well No. 16 s-s' (m) | Well No. 17 s-s' (m) | Δt (min) | Well No. 16 s-s' (m) | Well No. 17 s-s' (m) | Δt (min) | Well No. 16 s-s' (m) | Well No. 17 s-s' (m) |
|---|---|---|---|---|---|---|---|---|
| 0.17 | 0.00059 | 0.00118 | 7 | 0.01218 | 0.01762 | 70 | 0.02736 | 0.03138 |
| 0.33 | 0.00117 | 0.00261 | 8 | 0.01220 | 0.01922 | 80 | 0.02641 | 0.03185 |
| 0.5 | 0.00160 | 0.00476 | 9 | 0.01280 | 0.01824 | 90 | 0.02660 | 0.03333 |
| 0.67 | 0.00217 | 0.00605 | 10 | 0.01411 | 0.01898 | 100 | 0.02810 | 0.03326 |
| 0.83 | 0.00304 | 0.00734 | 12 | 0.01473 | 0.01903 | 120 | 0.03171 | 0.03458 |
| 1 | 0.00332 | 0.00791 | 14 | 0.01449 | 0.02108 | 150 | 0.03322 | 0.03938 |
| 1.33 | 0.00433 | 0.00963 | 16 | 0.01525 | 0.02041 | 180 | 0.03357 | 0.03916 |
| 1.67 | 0.00548 | 0.01021 | 18 | 0.01559 | 0.02232 | 210 | 0.03476 | 0.04006 |
| 2 | 0.00635 | 0.01065 | 20 | 0.01593 | 0.02065 | 240 | 0.03592 | 0.04222 |
| 2.5 | 0.00807 | 0.01152 | 25 | 0.01905 | 0.02293 | 270 | 0.03949 | 004493 |
| 3 | 0.00908 | 0.01267 | 30 | 0.02319 | 0.02363 | 300 | 0.03961 | 0.04577 |
| 3.5 | 0.00995 | 0.01382 | 35 | 0.01932 | 0.02605 | 330 | 0.04299 | 0.04858 |
| 4 | 0.01054 | 0.01412 | 40 | 0.02103 | 0.02605 | 360 | 0.04363 | 0.05007 |
| 5 | 0.01013 | 0.01543 | 50 | 0.02532 | 0.02819 | | | |
| 6 | 0.01001 | 0.01660 | 60 | 0.02391 | 0.03006 | | | |

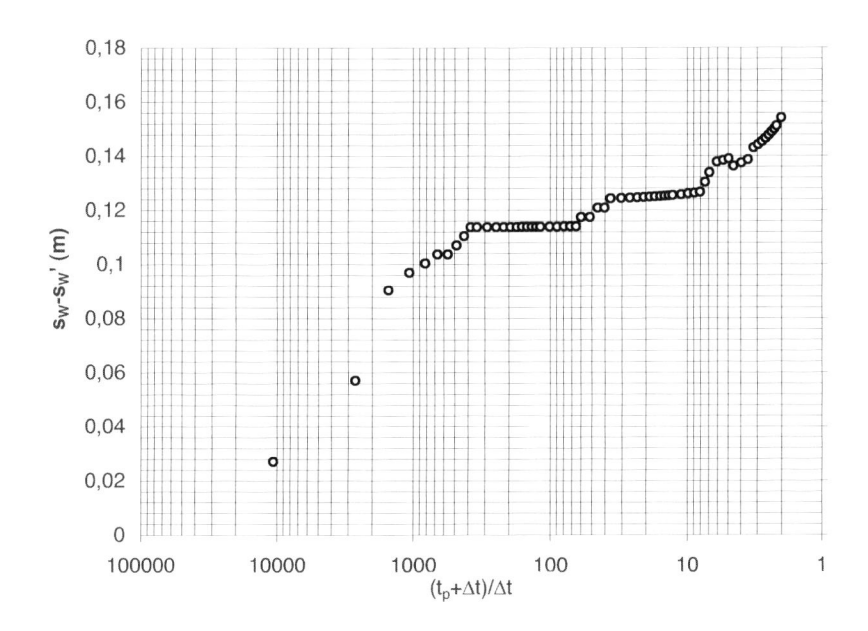

*Recovery data, pumped well*

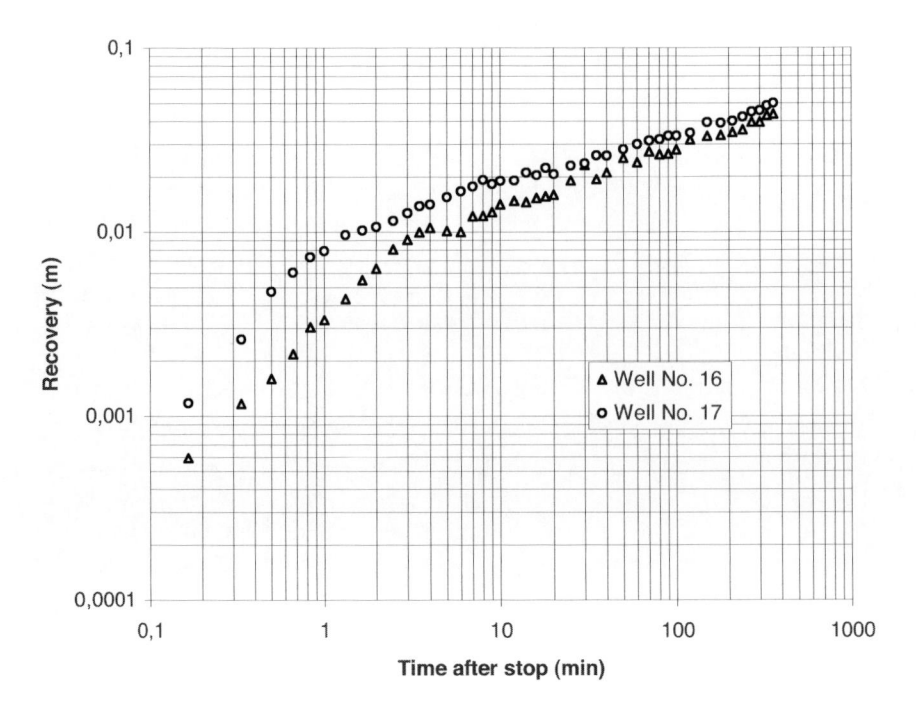

# FRACTURED AQUIFERS

*Recovery plot, monitoring wells*

Calculate hydraulic parameters. For the pumped well, carry out the skin calculation and write the equation for the recovery data.

**Problem 7.1**

Single fracture

*Gringarten and Witherspoon (1975)* published the recovery data below from an oil well. The well yield was $Q = 0.00097$ m³/s, $r_w = 0.076$ m, and the drawdown test duration was $t_p = 486{,}000$ min.

From a laboratory test, $S = 0.000207$. The drawdown at the end of the test was $s_w = 241.1$ m.

Determine the fracture type; calculate hydraulic parameters and skin using type-curve and semi-log methods.

| $\Delta t$ (min) | $(\Delta t + t_p)/\Delta t$ | $s_w - s_w'$ (m) |
|---|---|---|
| 5 | 93600 | 7.678 |
| 10 | 46700 | 10.47 |
| 15 | 31200 | 12.564 |
| 30 | 15600 | 17.101 |
| 45 | 10400 | 20.242 |
| 60 | 7800 | 22.336 |
| 120 | 3900 | 30.014 |
| 180 | 2600 | 35.598 |
| 240 | 1950 | 39.786 |
| 300 | 1560 | 43.276 |
| 360 | 1300 | 46.068 |
| 420 | 1120 | 48.86 |
| 480 | 976 | 52.35 |
| 540 | 868 | 54.444 |
| 600 | 781 | 55.84 |
| 720 | 651 | 60.028 |
| 1440 | 326 | 75.384 |
| 2160 | 218 | 86.552 |
| 2880 | 164 | 94.23 |
| 3600 | 131 | 99.814 |
| 4320 | 109 | 104.7 |
| 5760 | 82.3 | 113.076 |
| 7200 | 66 | 118.66 |
| 8640 | 55.2 | 125.64 |
| 11520 | 41.6 | 132.62 |
| 14400 | 33.5 | 139.6 |

# FRACTURED AQUIFERS

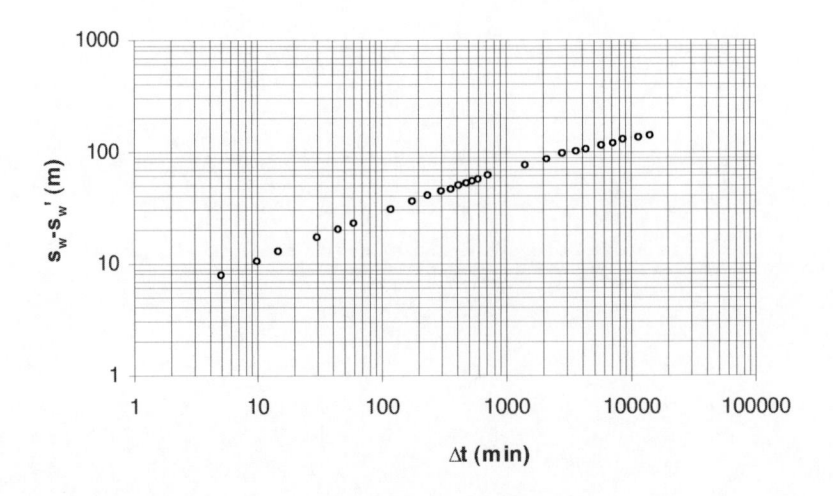

*Recovery data plot*

**Problem 7.2**

Single fracture

The table below shows drawdown data from an oil well published by *Cinco et al (1978)*. The well yield was $Q = 0.000502$ m³/s, $r_w = 0.0763$ m, and from laboratory tests $S = 0.000178$. The reservoir thickness is $h = 16.8$ m.

Determine the fracture type; calculate hydraulic parameters and skin using type-curve and semi-log methods.

| t (min) | s_w (m) | t (min) | s_w (m) |
|---|---|---|---|
| 60 | 55.809 | 1440 | 194.987 |
| 120 | 75.101 | 1800 | 211.523 |
| 180 | 88.192 | 2400 | 229.437 |
| 240 | 99.216 | 3000 | 245.284 |
| 300 | 108.173 | 3600 | 260.442 |
| 360 | 117.13 | 4200 | 272.155 |
| 420 | 125.398 | 4800 | 283.179 |
| 480 | 132.288 | 5400 | 292.136 |
| 540 | 138.489 | 6000 | 302.471 |
| 600 | 142.623 | 7200 | 316.251 |
| 720 | 153.647 | 9000 | 333.476 |
| 840 | 162.604 | 12000 | 359.658 |
| 960 | 170.183 | 15000 | 377.572 |
| 1200 | 183.963 | 18000 | 393.419 |

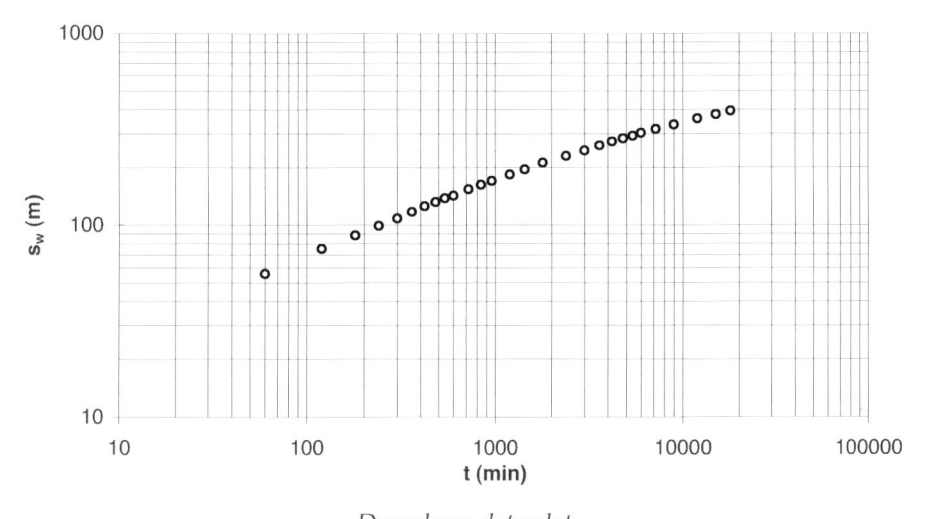

*Drawdown data plot*

**Problem 7.3**

Single fracture.

Data is from a confined fractured gneiss aquifer at Nybro, the Island of Bornholm, Denmark, *(Nielsen, 2007)*. The tables below show drawdown data from the pumped well and recovery data from two monitoring wells, No. 176 and No. 177 at distances 253 and 206 m from the pumped well. The tables give selected data, while the plots from the pumped well and the two monitoring wells show the full data set. The geological setting is a fracture system and from geological mapping, we can assume that all three wells are in the fracture system. The well yield was 8 m³/hr.

Prepare diagnostic plots and calculate hydraulic parameters and fracture geometry. Hint: After having determined the fracture type, use monitoring well type-curves with the same type of fracture.

### Pumped well

| t (min) | s_w (m) | t (min) | s_w (m) | t (min) | s_w (m) |
|---|---|---|---|---|---|
| 1 | 0.48 | 40 | 2.54 | 600 | 4.27 |
| 2 | 1.63 | 50 | 2.65 | 720 | 4.38 |
| 3 | 1.75 | 60 | 2.73 | 840 | 4.5 |
| 4 | 1.87 | 75 | 2.82 | 960 | 4.64 |
| 5 | 1.92 | 90 | 2.91 | 1080 | 4.77 |
| 6 | 1.99 | 105 | 2.99 | 1200 | 4.87 |
| 7 | 2.03 | 120 | 3.06 | 1400 | 5.02 |
| 8 | 2.08 | 150 | 3.18 | 1600 | 5.18 |
| 9 | 2.12 | 180 | 3.31 | 1800 | 5.31 |
| 10 | 2.15 | 210 | 3.4 | 2000 | 5.41 |
| 12 | 2.17 | 240 | 3.52 | 2500 | 5.7 |
| 14 | 2.22 | 270 | 3.57 | 3000 | 5.91 |
| 16 | 2.24 | 300 | 3.65 | 3500 | 6.06 |
| 18 | 2.27 | 360 | 3.8 | 4000 | 6.21 |
| 20 | 2.3 | 420 | 3.91 | 4500 | 6.29 |
| 25 | 2.37 | 480 | 4.03 | 5000 | 6.38 |
| 30 | 2.43 | 540 | 4.13 | 5500 | 6.45 |
|  |  |  |  | 5761 | 6.49 |

### Monitoring wells

|  | 177 | 176 |  | 177 | 176 |  | 177 | 176 |
|---|---|---|---|---|---|---|---|---|
| Δt (min) | s" (m) | s" (m) | Δt (min) | s" (m) | s" (m) | Δt (min) | s" (m) | s" (m) |
| 60 | 0.15 | 0.015 | 960 | 1.05 | 0.735 | 3360 | 1.785 | 1.395 |
| 120 | 0.29 | 0.06 | 1080 | 1.12 | 0.79 | 3600 | 1.855 | 1.44 |
| 180 | 0.395 | 0.125 | 1200 | 1.18 | 0.845 | 3960 | 1.955 | 1.49 |
| 240 | 0.48 | 0.19 | 1320 | 1.235 | 0.885 | 4320 | 2.035 | 1.53 |
| 300 | 0.57 | 0.265 | 1440 | 1.285 | 0.945 | 4680 | 2.095 | 1.575 |
| 360 | 0.635 | 0.32 | 1620 | 1.345 | 1 | 4920 | 2.12 | 1.595 |
| 420 | 0.685 | 0.37 | 1800 | 1.415 | 1.06 | 5400 | 2.16 | 1.645 |
| 480 | 0.74 | 0.435 | 1980 | 1.47 | 1.115 | 5880 | 2.22 | 1.68 |
| 540 | 0.78 | 0.475 | 2160 | 1.525 | 1.16 | 6240 | 2.28 | 1.725 |
| 600 | 0.83 | 0.52 | 2340 | 1.585 | 1.205 | 6960 | 2.34 | 1.78 |
| 660 | 0.88 | 0.56 | 2520 | 1.615 | 1.245 | 7380 | 2.38 | 1.8 |
| 720 | 0.93 | 0.6 | 2880 | 1.665 | 1.305 | 7860 | 2.43 | 1.845 |
| 840 | 0.99 | 0.665 | 3120 | 1.725 | 1.36 | 8520 | 2.47 | 1.87 |

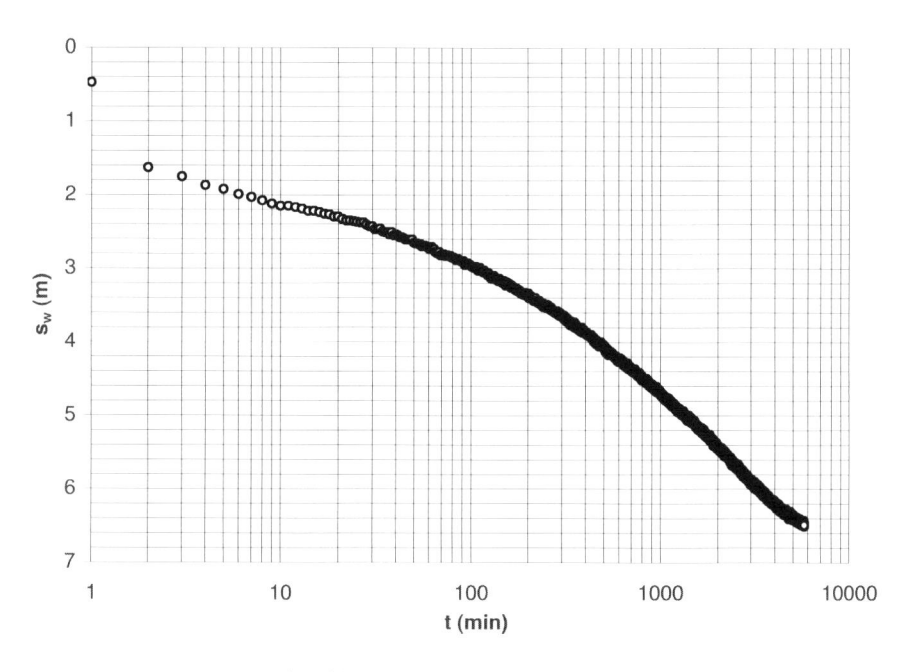

*Pumped well, semi-log drawdown data plot.*

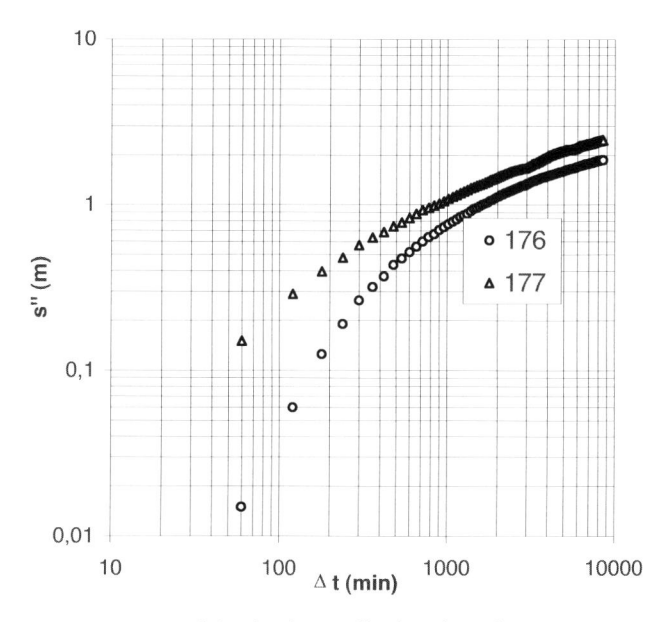

*Monitoring wells, log- log plot.*

205

FRACTURED AQUIFERS

## Problem 8.1

Single fracture

The table below shows drawdown data from pumped well MG-624 published by *Goode and Senior (1998)*. The well taps a fractured shale aquifer. The only information given is $Q = 0.00075$ m³/h. We assume the following: $r_w = 0.1$ m, aquifer thickness $h = 5$ m and $S = 0.0002$.

Determine the fracture type, hydraulic conductivities and fracture geometry.

| t (min) | $s_w$ (m) | t (min) | $s_w$ (m) |
|---|---|---|---|
| 0.5 | 0.96926 | 11 | 13.091 |
| 1 | 2.1549 | 12 | 13.746 |
| 1.5 | 2.8407 | 13 | 14.387 |
| 2.5 | 4.2642 | 14 | 15.127 |
| 3 | 4.8524 | 15 | 15.783 |
| 3.5 | 5.4315 | 16 | 16.435 |
| 4 | 6.0899 | 17 | 17.096 |
| 4.5 | 6.6599 | 18 | 17.508 |
| 5 | 7.3457 | 19 | 18.035 |
| 6 | 8.4521 | 22 | 19.513 |
| 7 | 9.5128 | 23 | 19.967 |
| 8 | 10.369 | 25 | 19.974 |
| 9 | 11.296 | | |

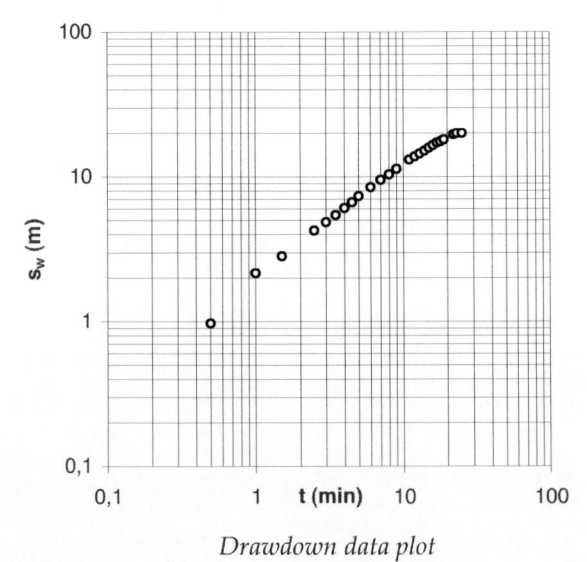

*Drawdown data plot*

206

## Problem 9.1

Combination of fracture systems

The table below shows drawdown data from a pumping well (Well No. 3585) and two monitoring wells (Well Nos. 79 and 2244) at Gjeddesdal, Denmark *(Kruger, 2000)*. The aquifer rock material is fractured limestone. The well yield was $Q = 88$ m³/hr, and $r_w = 0.1125$ m. The distance to monitoring wells 79 and 2244 are 510 m and 570 m, respectively. Other pumping in the area influences data and the monitoring wells are private wells in use during the test.

Prepare data diagnostics and determine hydraulic parameters for the pumped well and monitoring wells. Carry out skin calculation for the pumped well.

| Well No.3585 |        |         |          | Well No. 79 |        |         |       | Well No. 2244 |        |         |       |
|--------------|--------|---------|----------|-------------|--------|---------|-------|---------------|--------|---------|-------|
| t (min)      | $s_w$ (m) | t (min) | $s_w$(m) | t (min)     | s (m)  | t (min) | s (m) | t (min)       | s (m)  | t (min) | s (m) |
| 0.18         | 0.4046 | 100     | 1.204    | 28          | 0.06   | 15000   | 0.62  | 21            | 0.03   | 25000   | 0.71  |
| 0.33         | 0.658  | 120     | 1.207    | 42          | 0.091  | 20000   | 0.9   | 36            | 0.042  | 31000   | 0.75  |
| 0.42         | 0.71   | 150     | 1.237    | 60          | 0.115  | 25000   | 0.09  | 50            | 0.052  | 35000   | 0.78  |
| 0.51         | 0.741  | 200     | 1.261    | 75          | 0.12   | 30000   | 0.95  | 66            | 0.058  | 40000   | 0.82  |
| 0.71         | 0.755  | 250     | 1.301    | 90          | 0.135  | 35000   | 1.02  | 82            | 0.06   | 48000   | 0.79  |
| 0.9          | 0.778  | 300     | 1.306    | 100         | 0.14   | 40000   | 1.08  | 96            | 0.067  |         |       |
| 1.1          | 0.776  | 360     | 1.32     | 120         | 0.15   | 48000   | 1.1   | 120           | 0.07   |         |       |
| 1.3          | 0.776  | 420     | 1.329    | 150         | 0.16   |         |       | 150           | 0.075  |         |       |
| 1.5          | 0.776  | 480     | 1.329    | 180         | 0.17   |         |       | 180           | 0.083  |         |       |
| 1.8          | 0.776  | 540     | 1.349    | 210         | 0.17   |         |       | 210           | 0.088  |         |       |
| 2.2          | 0.776  | 600     | 1.352    | 240         | 0.175  |         |       | 250           | 0.09   |         |       |
| 2.5          | 0.776  | 800     | 1.349    | 300         | 0.18   |         |       | 330           | 0.095  |         |       |
| 3            | 0.793  | 1000    | 1.35     | 360         | 0.19   |         |       | 400           | 0.098  |         |       |
| 3.5          | 0.804  | 1200    | 1.36     | 450         | 0.2    |         |       | 500           | 0.098  |         |       |
| 4            | 0.804  | 1500    | 1.372    | 500         | 0.21   |         |       | 600           | 0.105  |         |       |
| 4.6          | 0.816  | 2100    | 1.384    | 600         | 0.22   |         |       | 700           | 0.11   |         |       |
| 5.5          | 0.828  | 2500    | 1.389    | 700         | 0.22   |         |       | 800           | 0.115  |         |       |
| 6.7          | 0.828  | 3200    | 1.397    | 800         | 0.225  |         |       | 900           | 0.12   |         |       |
| 7.5          | 0.83   | 4000    | 1.42     | 900         | 0.225  |         |       | 1000          | 0.125  |         |       |
| 8.8          | 0.852  | 5000    | 1.514    | 1000        | 0.23   |         |       | 1200          | 0.13   |         |       |
| 10.5         | 0.859  | 6000    | 1.56     | 1200        | 0.235  |         |       | 1500          | 0.14   |         |       |
| 12           | 0.871  | 7000    | 1.514    | 1500        | 0.26   |         |       | 1800          | 0.143  |         |       |
| 14           | 0.899  | 8000    | 1.514    | 1800        | 0.3    |         |       | 2100          | 0.15   |         |       |
| 16           | 0.899  | 9000    | 1.521    | 2100        | 0.32   |         |       | 2600          | 0.16   |         |       |

# FRACTURED AQUIFERS

| | | | | | | | | |
|---|---|---|---|---|---|---|---|---|
| 18 | 0.906 | 10000 | 1.573 | 2600 | 0.33 | | 3000 | 0.18 |
| 20 | 0.92 | 12000 | 1.656 | 3000 | 0.36 | | 3500 | 0.2 |
| 25 | 0.958 | 15000 | 1.632 | 3500 | 0.36 | | 4000 | 0.27 |
| 28 | 1.017 | 18000 | 1.704 | 4000 | 0.37 | | 5000 | 0.34 |
| 33 | 1.04 | 22000 | 1.704 | 5000 | 0.43 | | 6000 | 0.35 |
| 38 | 1.072 | 25000 | 1.751 | 6000 | 0.5 | | 7000 | 0.39 |
| 45 | 1.107 | 28000 | 1.751 | 7000 | 0.48 | | 8000 | 0.4 |
| 52 | 1.112 | 31000 | 1.798 | 8000 | 0.51 | | 9000 | 0.42 |
| 60 | 1.159 | 35000 | 1.822 | 9000 | 0.54 | | 10000 | 0.48 |
| 75 | 1.171 | 40000 | 1.869 | 10000 | 0.58 | | 12000 | 0.53 |
| 90 | 1.188 | 45000 | 1.751 | 12000 | 0.59 | | 15000 | 0.52 |
| | | | | | | | 20000 | 0.63 |

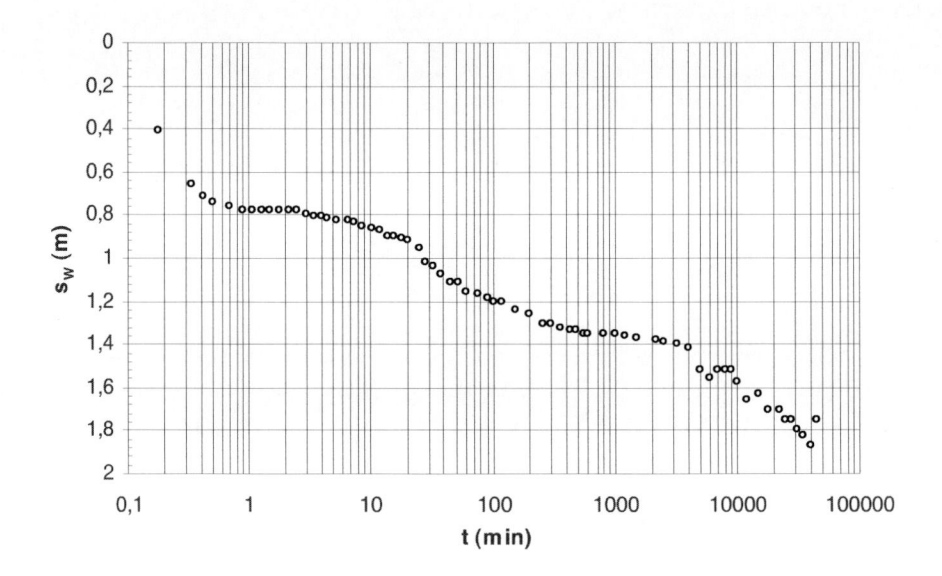

*Drawdown data, pumped well*

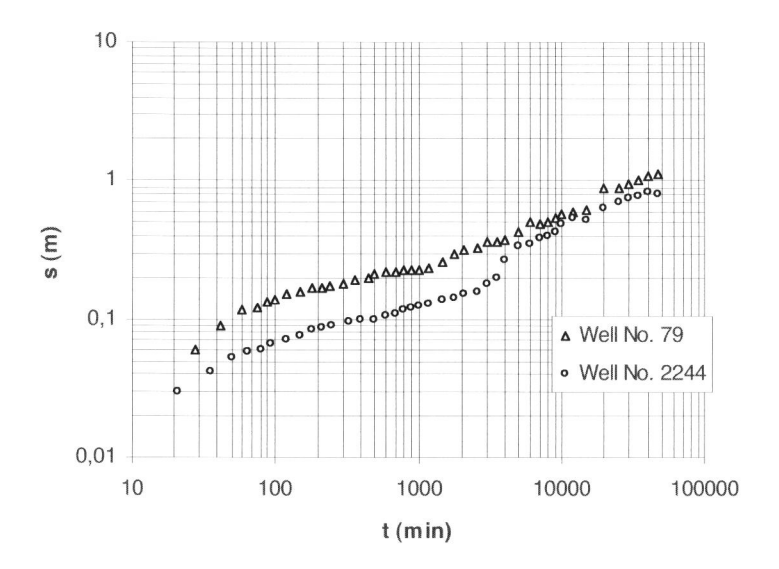

*Drawdown data, monitoring wells*

**Problem 11.1**

Linear discharge variation

Using Equation 11.2, show that Equation 11.14 is correct. Plot the type curve.

209

# Annex 2. Type-curve tables

Table 2.1 Streltsova, fractures

Table 2.2 Streltsova, matrix

Table 2.3 Najurieta, stratum case

Table 2.4 Najurieta, blocks case

Table 2.5 Gringarten et al, vertical fracture

Table 2.6 Gringarten and Witherspoon, vertical fracture, monitoring well on x-axis

Table 2.7 Gringarten and Witherspoon, vertical fracture, monitoring well on y-axis

Table 2.8 Gringarten and Witherspoon, vertical fracture, monitoring well on 45° line

Table 2.9 Cinco et al, finite hydraulic conductivity vertical fracture

Table 2.10 Gringarten and Ramey, horizontal fracture

### Table 2.1 Drawdown in fractures.
Streltsova (1976).

| t_D | 1/t_D | W(t_D) | ω=0.1 r/B=0.05 | ω=0.01 | ω=0.001 | ω=0.1 r/B=0.2 | ω=0.01 | ω=0.001 |
|---|---|---|---|---|---|---|---|---|
| 0.002 | 500 | | | 0.0011 | 0.5594 | | 0.0011 | 0.5533 |
| 0.005 | 200 | | | 0.0489 | 1.2209 | | 0.0487 | 1.1944 |
| 0.01 | 100 | | | 0.2193 | 1.8184 | | 0.2179 | 1.7528 |
| 0.02 | 50 | | 0.0011 | 0.5594 | 2.4576 | 0.0011 | 0.5534 | 2.3112 |
| 0.05 | 20 | | 0.0489 | 1..2209 | 3.3264 | 0.0487 | 1.1947 | 2.9526 |
| 0.1 | 10 | 4.24E-06 | 0.2193 | 1.8185 | 3.9796 | 0.2181 | 1.7534 | 3.2884 |
| 0.2 | 5 | 0.001148 | 0.5594 | 2.4577 | 4.6086 | 0.5539 | 2.3127 | 3.4585 |
| 0.5 | 2 | 0.0489 | 1.2211 | 3.3267 | 5.3541 | 1.1972 | 2.9566 | 3.509 |
| 1 | 1 | 0.2194 | 1.8189 | 3.9801 | 5.7971 | 1.7598 | 3.2967 | 3.5149 |
| 2 | 0.5 | 0.5598 | 2.4586 | 4.6096 | 6.833 | 2.3272 | 3.4753 | 3.5244 |
| 5 | 0.2 | 1.2227 | 3.3293 | 5.3569 | 6.2204 | 2.9961 | 3.5511 | 3.5525 |
| 10 | 0.1 | 1.8229 | 3.9894 | 5.8026 | 6.2344 | 3.3776 | 3.5983 | 3.5986 |
| 20 | 0.05 | 2.4679 | 4.6205 | 6.0945 | 6.2408 | 3.6359 | 3.673 | 3.6876 |
| 50 | 0.02 | 3.3547 | 5.3843 | 6.2481 | 6.2593 | 3.9277 | 3.9309 | 3.9311 |
| 100 | 0.01 | 4.0349 | 5.8574 | 6.2894 | 6.2897 | 4.2707 | 4.2721 | 4.2722 |
| 200 | 0.005 | 4.7261 | 6.2024 | 6.3491 | 6.3491 | 4.7816 | 4.7822 | 4.7823 |
| 500 | 0.002 | 5.6394 | 6.5055 | 6.5169 | 6.5169 | 5.6414 | 5.6414 | 5.6414 |
| 1000 | 0.001 | 6.3315 | 6.7653 | 6.7667 | 6.7657 | 6,3316 | 6,3316 | 6,3316 |
| 5000 | 0.0002 | 7.94 | 7.94 | 7.94 | 7.94 | 7.94 | 7.94 | 7.94 |

## Table 2.1 Continued

| $t_D$ | $1/t_D$ | $W(t_D)$ | $\omega=0.1$ r/B=0.5 | $\omega=0.01$ | $\omega=0.001$ | $\omega=0.1$ r/B=1 | $\omega=0.01$ | $\omega=0.001$ |
|---|---|---|---|---|---|---|---|---|
| 0.002 | 500 | | | 0.0011 | 0.5206 | | 0.0011 | 0.4212 |
| 0.005 | 200 | | | 0.0477 | 1.0593 | | 0.0445 | 0.7154 |
| 0.01 | 100 | | | 0.2104 | 1.4426 | | 0.1858 | 0.8204 |
| 0.02 | 50 | | 0.0011 | 0.521 | 1.7084 | 0.0011 | 0.4224 | 0.8438 |
| 0.05 | 20 | | 0.0478 | 1.0609 | 1.8403 | 0.0449 | 0.7204 | 0.8498 |
| 0.1 | 10 | 4.24E-06 | 0.2112 | 1.4463 | 1.8536 | 0.1888 | 0.82 | 0.8569 |
| 0.2 | 5 | 0.001148 | 0.5245 | 1.7167 | 1.859 | 0.4348 | 0.8688 | 0.8718 |
| 0.5 | 2 | 0.0489 | 1.0759 | 1.8624 | 1.8744 | 0.7696 | 0.9139 | 0.9156 |
| 1 | 1 | 0.2194 | 1.483 | 1.8985 | 1.8998 | 0.9453 | 0.9848 | 0.9865 |
| 2 | 0.5 | 0.5598 | 1.7978 | 1.9486 | 1.9495 | 1.1034 | 1.1187 | 1.1201 |
| 5 | 0.2 | 1.2227 | 2.071 | 2.0903 | 2.0911 | 1.4557 | 1.4647 | 1.4656 |
| 10 | 0.1 | 1.8229 | 2.2969 | 2.2934 | 2.3044 | 1.8998 | 1.9042 | 1.9046 |
| 20 | 0.05 | 2.4679 | 2.6569 | 2.661 | 2.6614 | 2.4837 | 2.485 | 2.485 |
| 50 | 0.02 | 3.3547 | 3.3791 | 3.3802 | 3.3803 | 3.3564 | 3.3566 | 3.3566 |
| 100 | 0.01 | 4.0349 | 4.0406 | 4.0408 | 4.0409 | 4.0384 | 4.0384 | 4.0384 |
| 200 | 0.005 | 4.7261 | 4.7265 | 4.7265 | 4.7265 | 4.7262 | 4.7262 | 4.7262 |
| 500 | 0.002 | 5.6394 | 5.6394 | 5.6394 | 5.6394 | 5.6249 | 5.6394 | 5.6394 |
| 1000 | 0.001 | 6.3315 | 6.3315 | 6.3315 | 6.3315 | 6.3315 | 6.3315 | 6.3315 |
| 5000 | 0.0002 | 7.94 | 7.94 | 7.94 | 7.94 | 7.94 | 7.94 | 7.94 |

## Table 2.1 Continued

| $t_D$ | $1/t_D$ | $W(t_D)$ | $\omega$=0.1 r/B=2 | $\omega$=0.01 | $\omega$=0.001 | $\omega$=0.1 r/B=3 | $\omega$=0.01 | $\omega$=0.001 |
|---|---|---|---|---|---|---|---|---|
| 0.002 | 500 | | | 0.0009 | 0.1946 | | 0.0006 | 0.0683 |
| 0.005 | 200 | | | 0.0335 | 0.2279 | | 0.021 | 0.0705 |
| 0.01 | 100 | | | 0.1148 | 0.2303 | | 0.0544 | 0.0719 |
| 0.02 | 50 | | 0.0009 | 0.1977 | 0.2331 | 0.0006 | 0.0718 | 0.0746 |
| 0.05 | 20 | | 0.0348 | 0.2385 | 0.2414 | 0.0231 | 0.0814 | 0.0828 |
| 0.1 | 10 | 4.24E-06 | 0.1233 | 0.2533 | 0.2553 | 0.0661 | 0.0953 | 0.0927 |
| 0.2 | 5 | 0.001148 | 0.2287 | 0.281 | 0.2829 | 0.1087 | 0.1236 | 0.1249 |
| 0.5 | 2 | 0.0489 | 0.3437 | 0.3622 | 0.3639 | 0.1991 | 0.2111 | 0.2123 |
| 1 | 1 | 0.2194 | 0.4771 | 0.4915 | 0.4959 | 0.3493 | 0.3587 | 0.3596 |
| 2 | 0.5 | 0.5598 | 0.7166 | 0.7264 | 0.7274 | 0.6287 | 0.6342 | 0.6347 |
| 5 | 0.2 | 1.2227 | 1.2638 | 1.2672 | 1.2676 | 1.2377 | 1.2392 | 1.2393 |
| 10 | 0.1 | 1.8229 | 1.8728 | 1.8339 | 1.834 | 1.8267 | 1.8271 | 1.8267 |
| 20 | 0.05 | 2.4679 | 2.4702 | 2.4702 | 2.4704 | 2.4687 | 2.4687 | 2.4687 |
| 50 | 0.02 | 3.3547 | 3.355 | 3.355 | 3.355 | 3.3547 | 3.3547 | 3.3547 |
| 100 | 0.01 | 4.0349 | 4.0379 | 4.0379 | 4.0379 | 4.0379 | 4.0379 | 4.0379 |
| 200 | 0.005 | 4.7261 | 4.7261 | 4.7261 | 4.7261 | 4.7261 | 4.7261 | 4.7261 |
| 500 | 0.002 | 5.6394 | 5.6394 | 5.6394 | 5.6394 | 5.6394 | 5.6394 | 5.6394 |
| 1000 | 0.001 | 6.3315 | 6.3315 | 6.3315 | 6.3315 | 6.3315 | 6.3315 | 6.3315 |
| 5000 | 0.0002 | 7.94 | 7.94 | 7.94 | 7.94 | 7.94 | 7.94 | 7.94 |

## Table 2.2 Drawdown in matrix.
Streltsova (1976).

| | | | ω=0.1 | ω=0.01 | ω=0.001 | ω=0.1 | ω=0.01 | ω=0.001 |
|---|---|---|---|---|---|---|---|---|
| $t_D$ | $1/t_D$ | $W(1/t_D)$ | r/B=0.05 | | | r/B=0.2 | | |
| 0.1 | 10 | 4.24E-06 | | | | | | |
| 0.2 | 5 | 0.001148 | | | | | | |
| 0.5 | 2 | 0.0489 | | | | | | |
| 1 | 1 | 0.2194 | | | | | | |
| 2 | 0.5 | 0.5598 | 0.00004 | 0.00028 | 0.00076 | 0.00045 | 0.0036 | 0.0071 |
| 5 | 0.2 | 1.2227 | 0.0002 | 0.00084 | 0.0019 | 0.0026 | 0.0111 | 0.0175 |
| 10 | 0.1 | 1.8229 | 0.0006 | 0.0019 | 0.0039 | 0.0093 | 0.0267 | 0.034 |
| 20 | 0.05 | 2.4679 | 0.0018 | 0.0045 | 0.0078 | 0.028 | 0.059 | 0.0695 |
| 50 | 0.02 | 3.3547 | 0.007 | 0.0138 | 0.0194 | 0.1019 | 0.1622 | 0.172 |
| 100 | 0.01 | 4.0349 | 0.0178 | 0.031 | 0.0388 | 0.2444 | 0.3279 | 0.3379 |
| 200 | 0.005 | 4.7261 | 0.0434 | 0.0676 | 0.0775 | 0.5382 | 0.6419 | 0.6523 |
| 500 | 0.002 | 5.6394 | 0.1329 | 0.18 | 0.1921 | 1.3326 | 1.4609 | 1.4722 |
| 1000 | 0.001 | 6.3315 | 0.2963 | 0.365 | 0.3793 | 2.3494 | 2.5014 | 2.5138 |
| 2000 | 0.0005 | 7.0242 | 0.6327 | 0.7209 | 0.7392 | 3.6318 | 3.8076 | 3.8211 |
| 5000 | 0.0002 | 7.9402 | 1.5807 | 1.6845 | 1.7133 | 5.1574 | 5.3476 | 5.3618 |
| 10000 | 0.0001 | 8.6332 | 2.8227 | 2.9959 | 3.0385 | 6.0105 | 6.2019 | 6.2162 |
| 20000 | 0.00005 | 9.3263 | 4.6957 | 4.8189 | 4.8789 | 6.7631 | 6.9546 | 6.969 |

| | | | ω=0.1 | ω=0.01 | ω=0.001 | ω=0.1 | ω=0.01 | ω=0.001 |
|---|---|---|---|---|---|---|---|---|
| $t_D$ | $1/t_D$ | $W(1/t_D)$ | r/B=0.5 | | | r/B=1 | | |
| 0.1 | 10 | 4.24E-06 | | | | | | |
| 0.2 | 5 | 0.001148 | | | | | | |
| 0.5 | 2 | 0.0489 | | | | | | |
| 1 | 1 | 0.2194 | | | | | | |
| 2 | 0.5 | 0.5598 | 0.0027 | 0.018 | 0.0235 | | | |
| 5 | 0.2 | 1.2227 | 0.0149 | 0.0489 | 0.0573 | 0.0439 | 0.0963 | 0.1034 |
| 10 | 0.1 | 1.8229 | 0.0498 | 0.1049 | 0.1136 | 0.1262 | 0.195 | 0.2032 |
| 20 | 0.05 | 2.4679 | 0.1376 | 0.2142 | 0.2233 | 0.296 | 0.3821 | 0.3921 |
| 50 | 0.02 | 3.3547 | 0.4202 | 0.5209 | 0.5311 | 0.7481 | 0.8682 | 0.8821 |
| 100 | 0.01 | 4.0349 | 0.8478 | 0.9689 | 0.9805 | 1.3264 | 1.4777 | 1.4962 |
| 200 | 0.005 | 4.7261 | 1.5293 | 1.6768 | 1.6904 | 2.064 | 2.2382 | 2.258 |
| 500 | 0.002 | 5.6394 | 2.738 | 2.9167 | 2.9328 | 3.0651 | 3.2505 | 3.2715 |
| 1000 | 0.001 | 6.3315 | 3.6406 | 3.8286 | 3.8455 | 3.7882 | 3.9762 | 3.9974 |
| 2000 | 0.0005 | 7.0242 | 4.4319 | 4.6221 | 4.6391 | 4.4956 | 4.6848 | 4.7061 |
| 5000 | 0.0002 | 7.9402 | 5.3981 | 5.589 | 5.6062 | 5.4203 | 5.6102 | 5.6316 |
| 10000 | 0.0001 | 8.6332 | 6.1067 | 6.2979 | 6.3151 | 6.1163 | 6.3063 | 6.3277 |
| 20000 | 0.00005 | 9.3263 | 6.8074 | 6.9987 | 7.0159 | 6.8168 | 7.001 | 7.0224 |

## Table 2.2 Continued.

| | | | $\omega=0.1$ | $\omega=0.01$ | $\omega=0.001$ | $\omega=0.1$ | $\omega=0.01$ | $\omega=0.001$ |
|---|---|---|---|---|---|---|---|---|
| $t_D$ | $1/t_D$ | $W(1/t_D)$ | $r/B=2$ | | | $r/B=3$ | | |
| 0.1 | 10 | 4.24E-06 | | | | | | |
| 0.2 | 5 | 0.001148 | | | | | | |
| 0.5 | 2 | 0.0489 | | | | | | |
| 1 | 1 | 0.2194 | | | | | | |
| 2 | 0.5 | 0.5598 | | | | | | |
| 5 | 0.2 | 1.2227 | 0.0703 | 0.1137 | 0.1181 | 0.0749 | 0.135 | 0.14 |
| 10 | 0.1 | 1.8229 | 0.1735 | 0.2368 | 0.2419 | 0.183 | 0.27 | 0.27 |
| 20 | 0.05 | 2.4679 | 0.3855 | 0.455 | 0.4873 | 0.4213 | 0.504 | 0.5134 |
| 50 | 0.02 | 3.3547 | 0.9422 | 1.0837 | 1.0973 | 1.0167 | 1.1485 | 1.1628 |
| 100 | 0.01 | 4.0349 | 1.5519 | 1.7188 | 1.735 | 1.6146 | 1.7676 | 1.7841 |
| 200 | 0.005 | 4.7261 | 2.2221 | 2.4014 | 2.4187 | 2.2645 | 2.4283 | 2.4457 |
| 500 | 0.002 | 5.6394 | 3.1289 | 3.3161 | 3.3331 | 3.156 | 3.3263 | 3.3444 |
| 1000 | 0.001 | 6.3315 | 3.8194 | 4.0078 | 4.026 | 3.8411 | 4.0135 | 4.0318 |
| 2000 | 0.0005 | 7.0242 | 4.5112 | 4.7007 | 4.719 | 4.5303 | 4.7037 | 4.7222 |
| 5000 | 0.0002 | 7.9402 | 5.4268 | 5.6169 | 5.6353 | 5.4442 | 5.6183 | 5.6368 |
| 10000 | 0.0001 | 8.6332 | 6.1197 | 6.3101 | 6.3285 | 6.1365 | 6.3108 | 6.3294 |
| 20000 | 0.00005 | 9.3263 | 6.8207 | 7.0032 | 7.0216 | 6.8292 | 7.0037 | 7.0222 |

## Table 2.3 Najurieta drawdown in fractures for $\phi = 0$, stratum case.
Sorensen (1981).

| $t_D$ | $\Psi = 0$ | 0.5 | 1 | 2 | 5 | 10 | 20 | 50 |
|---|---|---|---|---|---|---|---|---|
| 0.01 | | 0.000001 | 0.000004 | 0.000111 | 0.002151 | 0.0106 | 0.03535 | 0.1146 |
| 0.0125 | | 2.1E-06 | 1.33E-05 | 0.000247 | 0.003779 | 0.0162 | 0.0489 | 0.144 |
| 0.016667 | | 3.1E-06 | 5.02E-05 | 0.00066 | 0.00734 | 0.02566 | 0.0716 | 0.1865 |
| 0.025 | | 1.33E-05 | 0.000247 | 0.00216 | 0.0165 | 0.0489 | 0.1146 | 0.2631 |
| 0.05 | | 0.000256 | 0.00216 | 0.0106 | 0.0489 | 0.1146 | 0.2194 | 0.4242 |
| 0.066667 | | 0.000695 | 0.00443 | 0.0183 | 0.0713 | 0.1538 | 0.2751 | 0.5056 |
| 0.1 | 4.15E-06 | 0.00254 | 0.0109 | 0.0354 | 0.1146 | 0.2194 | 0.3714 | 0.627 |
| 0.125 | 3.77E-05 | 0.00448 | 0.0166 | 0.0489 | 0.144 | 0.2629 | 0.4224 | 0.7024 |
| 0.166667 | 0.00036 | 0.00954 | 0.02809 | 0.0716 | 0.187 | 0.325 | 0.506 | 0.806 |
| 0.2 | 0.001148 | 0.0146 | 0.0383 | 0.089 | 0.2194 | 0.3714 | 0.5598 | 0.868 |
| 0.25 | 0.003779 | 0.0246 | 0.0541 | 0.1162 | 0.2625 | 0.4224 | 0.627 | 0.949 |
| 0.333333 | 0.01305 | 0.0483 | 0.0828 | 0.1561 | 0.3257 | 0.506 | 0.7248 | 1.058 |
| 0.5 | 0.0489 | 0.0972 | 0.143 | 0.2323 | 0.428 | 0.627 | 0.868 | 1.2227 |
| 0.666667 | 0.1 | 0.1561 | 0.209 | 0.302 | 0.5109 | 0.7248 | 0.973 | 1.3431 |
| 1 | 0.2194 | 0.2785 | 0.3329 | 0.43 | 0.644 | 0.8698 | 1.131 | 1.518 |
| 1.25 | 0.3106 | 0.3682 | 0.416 | 0.5151 | 0.7306 | 0.955 | 1.2227 | 1.602 |
| 1.666667 | 0.4544 | 0.50764 | 0.555 | 0.642 | 0.847 | 1.073 | 1.3431 | 1.737 |
| 2.5 | 0.7024 | 0.746 | 0.786 | 0.861 | 1.0411 | 1.238 | 1.522 | 1.9246 |
| 5 | 1.2227 | 1.2398 | 1.276 | 1.323 | 1.453 | 1.617 | 1.8563 | 2.2494 |
| 6.666667 | 1.4645 | 1.488 | 1.506 | 1.543 | 1.652 | 1.797 | 2.0122 | 2.387 |
| 10 | 1.8229 | 1.838 | 1.851 | 1.881 | 1.949 | 2.07 | 2.252 | 2.593 |
| 12.5 | 2.0269 | 2.0386 | 2.051 | 2.0744 | 2.141 | 2.237 | 2.3966 | 2.715 |
| 16.66667 | 2.2953 | 2.305 | 2.313 | 2.332 | 2.384 | 2.461 | 2.591 | 2.881 |
| 25 | 2.6813 | 2.685 | 2.696 | 2.7056 | 2.743 | 2.804 | 2.8996 | 3.132 |
| 50 | 3.3547 | 3.36 | 3.36 | 3.337 | 3.375 | 3.273 | 3.447 | 3.624 |
| 66.66667 | 3.6374 | 3.638 | 3.642 | 3.644 | 3.671 | 3.681 | 3.727 | 3.849 |
| 100 | 4.0379 | 4.0379 | 4.044 | 4.044 | 4.0539 | 4.071 | 4.101 | 4.185 |
| 125 | 4.2591 | 4.2591 | 4.2591 | 4.264 | 4.2716 | 4.285 | 4.3231 | 4.338 |
| 166.6667 | 4.5448 | 4.5448 | 4.5448 | 4.548 | 4.554 | 4.5649 | 4.583 | 4.6387 |
| 250 | 4.9482 | 4.9482 | 4.9482 | 4.94 | 4.953 | 4.9607 | 4.9735 | 5.013 |
| 500 | 5.6394 | 5.6394 | 5.6394 | 5.6394 | 5.645 | 5.644 | 5.675 | 5.67 |
| 666.6667 | 5.9266 | 5.9266 | 5.9266 | 5.9266 | 5.9266 | 5.935 | 5.94 | 5.96 |
| 1000 | 6.3315 | 6.3315 | 6.3315 | 6.3315 | 6.3315 | 6.338 | 6.34 | 6.348 |
| 1250 | 6.5545 | 6.5545 | 6.5545 | 6.5545 | 6.5545 | 6.5532 | 6.56 | 6.5545 |
| 1666.667 | 6.842 | 6.842 | 6.842 | 6.842 | 6.842 | 6.842 | 6.85 | 6.852 |
| 2500 | 7.2472 | 7.2472 | 7.2472 | 7.2472 | 7.2472 | 7.2472 | 7.25 | 7.254 |
| 5000 | 7.9402 | 7.9402 | 7.9402 | 7.9402 | 7.9402 | 7.9402 | 7.945 | 7.946 |
| 6666.667 | 8.2278 | 8.2278 | 8.2278 | 8.2278 | 8.2278 | 8.2278 | 8.228 | 8.23 |
| 10000 | 8.6332 | 8.6332 | 8.6332 | 8.6332 | 8.6332 | 8.6332 | 8.6332 | 8.635 |

## Table 2.4 Najurieta drawdown in fractures for $\phi = 0$, blocks case.

| $t_D$ | $\varepsilon = 0$ | $\varepsilon = 0.5$ | $\varepsilon = 1$ | $\varepsilon = 2$ | $\varepsilon = 5$ | $\varepsilon = 10$ | $\varepsilon = 20$ | $\varepsilon = 50$ |
|---|---|---|---|---|---|---|---|---|
| 0.01 | | 7.5E-09 | 1E-09 | 6E-08 | 0.000001 | 1.13E-05 | 0.000191 | 0.003033 |
| 0.0125 | | 6.5E-09 | 2E-08 | 1E-07 | 0.000002 | 3.15E-05 | 0.000429 | 0.005228 |
| 0.016667 | | 6.5E-08 | 6E-08 | 0.000001 | 5.22E-06 | 0.00011 | 0.00109 | 0.00982 |
| 0.025 | | 6E-08 | 5E-07 | 0.000002 | 4.29E-05 | 0.000509 | 0.003383 | 0.02124 |
| 0.05 | | 0.000001 | 0.000004 | 3.83E-05 | 0.00072 | 0.004057 | 0.01582 | 0.06156 |
| 0.066667 | | 0.000004 | 1.87E-05 | 0.000159 | 0.00185 | 0.008163 | 0.02675 | 0.08863 |
| 0.1 | 4.15E-06 | 4.81E-05 | 0.000192 | 0.000918 | 0.005861 | 0.01918 | 0.05102 | 0.1397 |
| 0.125 | 3.77E-05 | 0.000214 | 0.000598 | 0.002064 | 0.01021 | 0.02913 | 0.06998 | 0.175 |
| 0.166667 | 0.00036 | 0.001102 | 0.0021 | 0.00541 | 0.01956 | 0.04875 | 0.1011 | 0.2272 |
| 0.2 | 0.001148 | 0.002558 | 0.00451 | 0.00941 | 0.02844 | 0.06292 | 0.1259 | 0.2653 |
| 0.25 | 0.003779 | 0.00656 | 0.00989 | 0.0175 | 0.04367 | 0.0867 | 0.1609 | 0.3172 |
| 0.333333 | 0.01305 | 0.01833 | 0.024 | 0.0354 | 0.07209 | 0.1272 | 0.2157 | 0.3923 |
| 0.5 | 0.0489 | 0.05686 | 0.0687 | 0.0849 | 0.1348 | 0.2057 | 0.3112 | 0.514 |
| 0.666667 | 0.1 | 0.1108 | 0.1272 | 0.1433 | 0.2004 | 0.2776 | 0.3946 | 0.612 |
| 1 | 0.2194 | 0.2316 | 0.2434 | 0.265 | 0.3265 | 0.4098 | 0.5351 | 0.7662 |
| 1.25 | 0.3106 | 0.3227 | 0.3323 | 0.3566 | 0.4161 | 0.4988 | 0.6248 | 0.8597 |
| 1.666667 | 0.4544 | 0.4645 | 0.476 | 0.4966 | 0.5526 | 0.6312 | 0.7543 | 0.989 |
| 2.5 | 0.7024 | 0.7109 | 0.719 | 0.7347 | 0.7842 | 0.853 | 0.9649 | 1.1897 |
| 5 | 1.2227 | 1.226 | 1.255 | 1.244 | 1.2746 | 1.3208 | 1.4035 | 1.5863 |
| 6.666667 | 1.4645 | 1.465 | 1.468 | 1.484 | 1.5068 | 1.5444 | 1.6129 | 1.7749 |
| 10 | 1.8229 | 1.822 | 1.829 | 1.835 | 1.8521 | 1.88 | 1.9313 | 2.0592 |
| 12.5 | 2.0269 | 2.026 | 2.031 | 2.0368 | 2.0512 | 2.074 | 2.1172 | 2.2278 |
| 16.66667 | 2.2953 | 2.2953 | 2.299 | 2.3029 | 2.3185 | 2.3317 | 2.3657 | 2.4668 |
| 25 | 2.6813 | 2.6813 | 2.684 | 2.686 | 2.6948 | 2.705 | 2.7302 | 2.796 |
| 50 | 3.3547 | 3.3547 | 3.3547 | 3.3547 | 3.3617 | 3.3678 | 3.3809 | 3.4167 |
| 66.66667 | 3.6374 | 3.6374 | 3.6374 | 3.6374 | 3.6428 | 3.6476 | 3.6578 | 3.685 |
| 100 | 4.0379 | 4.0379 | 4.0379 | 4.0379 | 4.0413 | 4.0446 | 4.051 | 4.0701 |
| 125 | 4.2591 | 4.2591 | 4.2591 | 4.2591 | 4.2619 | 4.2644 | 4.2696 | 4.2862 |
| 166.6667 | 4.5448 | 4.5448 | 4.5448 | 4.5448 | 4.5468 | 4.5488 | 4.5528 | 4.5569 |
| 250 | 4.9482 | 4.9482 | 4.9482 | 4.9482 | 4.9482 | 4.9513 | 4.9535 | 4.9618 |
| 500 | 5.6394 | 5.6394 | 5.6394 | 5.6394 | 5.6394 | 5.6394 | 5.639 | 5.6466 |
| 666.6667 | 5.9266 | 5.9266 | 5.9266 | 5.9266 | 5.9266 | 5.9266 | 5.9266 | 5.9321 |
| 1000 | 6.3315 | 6.3315 | 6.3315 | 6.3315 | 6.3315 | 6.3315 | 6.3315 | 6.3348 |
| 1250 | 6.5545 | 6.5545 | 6.5545 | 6.5545 | 6.5545 | 6.5545 | 6.5545 | 6.5574 |
| 1666.667 | 6.842 | 6.842 | 6.842 | 6.842 | 6.842 | 6.842 | 6.842 | 6.844 |
| 2500 | 7.2472 | 7.2472 | 7.2472 | 7.2472 | 7.2472 | 7.2472 | 7.2472 | 7.247 |
| 5000 | 7.9402 | 7.9402 | 7.9402 | 7.9402 | 7.9402 | 7.9402 | 7.9402 | 7.9402 |
| 6666.667 | 8.2278 | 8.2278 | 8.2278 | 8.2278 | 8.2278 | 8.2278 | 8.2278 | 8.2278 |
| 10000 | 8.6332 | 8.6332 | 8.6332 | 8.6332 | 8.6332 | 8.6332 | 8.6332 | 8.6332 |

## Table 2.5 Drawdown in a well in a vertical fracture.
Gringarten, Raghaven and Ramey (1974).

| $t_D$ | Uniform flux | Infinite hydrailic conductivity | $t_D$ | Uniform flux | Infinite hydrailic conductivity |
|---|---|---|---|---|---|
| 0.0001 | 0.0354 | 0.0354 | 1 | 2.8894 | 2.406 |
| 0.00015 | 0.0434 | 0.0434 | 1.5 | 3.269 | 2.7508 |
| 0.0002 | 0.0501 | 0.0501 | 2 | 3.543 | 3.006 |
| 0.0003 | 0.0614 | 0.0614 | 3 | 3.935 | 3.378 |
| 0.0004 | 0.0709 | 0.0709 | 4 | 4.216 | 3.65 |
| 0.0005 | 0.0793 | 0.0793 | 5 | 4.435 | 3.862 |
| 0.0006 | 0.0868 | 0.0868 | 6 | 4.615 | 4.038 |
| 0.0008 | 0.1003 | 0.1003 | 8 | 4.899 | 4.316 |
| 0.001 | 0.1121 | 0.1121 | 10 | 5.12 | 4.534 |
| 0.0015 | 0.1373 | 0.1373 | 15 | 5.5226 | 4.932 |
| 0.002 | 0.1585 | 0.1585 | 20 | 5.809 | 5.218 |
| 0.003 | 0.1942 | 0.1942 | 30 | 6.213 | 5.618 |
| 0.004 | 0.2242 | 0.2242 | 40 | 6.5 | 5.904 |
| 0.005 | 0.2507 | 0.2507 | 50 | 6.722 | 6.126 |
| 0.006 | 0.2746 | 0.2746 | 60 | 6.905 | 6.308 |
| 0.008 | 0.3171 | 0.3171 | 80 | 7.192 | 6.596 |
| 0.01 | 0.3544 | 0.353 | 100 | 7.415 | 6.818 |
| 0.015 | 0.4342 | 0.429 | 150 | 7.8202 | 7.222 |
| 0.02 | 0.5014 | 0.4912 | 200 | 8.1078 | 7.51 |
| 0.03 | 0.614 | 0.591 | 300 | 8.5132 | 7.916 |
| 0.04 | 0.709 | 0.6712 | 400 | 8.8 | 8.202 |
| 0.05 | 0.7926 | 0.7394 | 500 | 9.024 | 8.4254 |
| 0.06 | 0.868 | 0.7992 | 600 | 9.206 | 8.6078 |
| 0.08 | 1.0014 | 0.9018 | 800 | 9.494 | 8.8954 |
| 0.1 | 1.1174 | 0.989 | 1000 | 9.717 | 9.118 |
| 0.15 | 1.358 | 1.1666 | 1500 | 10.122 | 9.524 |
| 0.2 | 1.5512 | 1.3098 | 2000 | 10.41 | 9.812 |
| 0.3 | 1.8522 | 1.5382 | 3000 | 10.815 | 10.216 |
| 0.4 | 2.0834 | 1.7206 | 4000 | 11.103 | 10.504 |
| 0.5 | 2.271 | 1.8734 | 5000 | 11.33 | 10.728 |
| 0.6 | 2.429 | 2.0054 | 6000 | 11.51 | 10.91 |
| 0.8 | 2.6854 | 2.2258 | 8000 | 11.8 | 11.1976 |

## Table 2.6 Vertical fracture. Drawdown in a monitoring well on x-axis.
Kruseman and de Ridder (2001).

| t_D/r' | r'=0.2 | 0.4 | 0.6 | 0.8 | 1.05 | 1.2 | 1.5 | 3 | 5 |
|---|---|---|---|---|---|---|---|---|---|
| 0.001 | 0.05013 | 0.0709 | 0.0868 | 0.1 | 0.00434 | | | | |
| 0.0015 | 0.0614 | 0.0868 | 0.106 | 0.123 | 0.00854 | | | | |
| 0.002 | 0.0709 | 0.1 | 0.122 | 0.1418 | 0.0129 | 0.00003 | | | |
| 0.003 | 0.08683 | 0.1228 | 0.15 | 0.1736 | 0.0219 | 0.00022 | | | |
| 0.004 | 0.10027 | 0.1418 | 0.173 | 0.2 | 0.0305 | 0.0007 | | | |
| 0.006 | 0.1228 | 0.1737 | 0.212 | 0.245 | 0.0465 | 0.00255 | | | |
| 0.008 | 0.1418 | 0.2005 | 0.245 | 0.282 | 0.0613 | 0.00539 | 0.00002 | | |
| 0.01 | 0.1585 | 0.2242 | 0.275 | 0.3137 | 0.075 | 0.00894 | 0.00007 | | |
| 0.015 | 0.1941 | 0.2745 | 0.336 | 0.3794 | 0.1058 | 0.01964 | 0.00056 | | |
| 0.02 | 0.2242 | 0.3171 | 0.388 | 0.4325 | 0.133 | 0.03168 | 0.00176 | | |
| 0.03 | 0.2746 | 0.3883 | 0.474 | 0.5175 | 0.181 | 0.05703 | 0.00639 | | |
| 0.04 | 0.3171 | 0.4484 | 0.546 | 0.5856 | 0.222 | 0.08244 | 0.01348 | | |
| 0.06 | 0.3883 | 0.549 | 0.663 | 0.694 | 0.294 | 0.1313 | 0.03243 | 0.00004 | |
| 0.08 | 0.4484 | 0.6334 | 0.759 | 0.7814 | 0.3555 | 0.1773 | 0.05494 | 0.00029 | |
| 0.1 | 0.5013 | 0.7074 | 0.84 | 0.8557 | 0.4106 | 0.22045 | 0.0792 | 0.00096 | |
| 0.15 | 0.614 | 0.8622 | 1 | 1.01 | 0.5295 | 0.3188 | 0.1426 | 0.00554 | 0.00012 |
| 0.2 | 0.7088 | 0.9892 | 1.139 | 1.13 | 0.631 | 0.4067 | 0.206 | 0.01485 | 0.00073 |
| 0.3 | 0.8672 | 1.194 | 1.351 | 1.329 | 0.802 | 0.5608 | 0.3274 | 0.04568 | 0.00528 |
| 0.4 | 0.999 | 1.359 | 1.52 | 1.489 | 0.947 | 0.694 | 0.4395 | 0.0874 | 0.0157 |
| 0.6 | 1.216 | 1.62 | 1.785 | 1.745 | 1.183 | 0.917 | 0.6369 | 0.1853 | 0.05246 |
| 0.8 | 1.391 | 1.823 | 1.992 | 1.947 | 1.3739 | 1.1 | 0.8046 | 0.287 | 0.10295 |
| 1 | 1.539 | 1.992 | 2.163 | 2.114 | 1.534 | 1.256 | 0.9496 | 0.3854 | 0.16 |
| 1.5 | 1.8345 | 2.313 | 2.49 | 2.439 | 1.849 | 1.564 | 1.2425 | 0.6059 | 0.3081 |
| 2 | 2.062 | 2.563 | 2.74 | 2.685 | 2.09 | 1.799 | 1.47 | 0.7919 | 0.4484 |
| 3 | 2.402 | 2.924 | 3.1 | 3.045 | 2.444 | 2.1506 | 1.812 | 1.089 | 0.6914 |
| 4 | 2.656 | 3.1886 | 3.369 | 3.31 | 2.71 | 2.41 | 2.0669 | 1.319 | 0.8915 |
| 6 | 3.0211 | 3.57 | 3.75 | 3.69 | 3.084 | 2.786 | 2.4387 | 1.665 | 1.2047 |
| 8 | 3.296 | 3.846 | 4.02 | 3.97 | 3.358 | 3.06 | 2.7092 | 1.922 | 1.4444 |
| 10 | 3.5077 | 4.061 | 4.24 | 4.18 | 3.573 | 3.273 | 2.922 | 2.127 | 1.638 |
| 15 | 3.8983 | 4.457 | 4.64 | 4.578 | 3.968 | 3.667 | 3.313 | 2.508 | 2.004 |
| 20 | 4.1785 | 4.74 | 4.92 | 4.86 | 4.25 | 3.95 | 3.594 | 2.783 | 2.271 |
| 30 | 4.5764 | 5.14 | 5.32 | 5.26 | 4.65 | 4.35 | 3.992 | 3.175 | 2.656 |
| 40 | 4.8605 | 5.426 | 5.61 | 5.546 | 4.93 | 4.63 | 4.276 | 3.457 | 2.933 |
| 60 | 5.2623 | 5.829 | 6.01 | 5.95 | 5.338 | 5.037 | 4.679 | 3.855 | 3.329 |
| 80 | 5.5482 | 6.116 | 6.3 | 6.245 | 5.624 | 5.324 | 4.966 | 4.14 | 3.61 |
| 100 | 5.77 | 6.338 | 6.52 | 6.459 | 5.847 | 5.546 | 5.189 | 4.363 | 3.83 |
| 150 | 6.174 | 6.743 | 6.93 | 6.863 | 6.252 | 5.9515 | 5.5942 | 4.764 | 4.231 |
| 200 | 6.462 | 7.031 | 7.21 | 7.15 | 6.539 | 6.239 | 5.882 | 5.05 | 4.517 |
| 300 | 6.867 | 7.436 | 7.62 | 7.555 | 6.944 | 6.644 | 6.2879 | 5.455 | 4.92 |

**Table 2.6 continued.**

| $t_D/r'$ | $r'=0.2$ | 0.4 | 0.6 | 0.8 | 1.05 | 1.2 | 1.5 | 3 | 5 |
|---|---|---|---|---|---|---|---|---|---|
| 400 | 7.155 | 7.723 | 7.905 | 7.843 | 7.232 | 6.9325 | 6.576 | 5.744 | 5.207 |
| 600 | 7.56 | 8.128 | 8.31 | 8.249 | 7.638 | 7.339 | 6.983 | 6.151 | 5.612 |
| 800 | 7.849 | 8.415 | 8.6 | 8.537 | 7.927 | 7.627 | 7.27 | 6.444 | 5.899 |
| 1000 | 8.0715 | 8.6375 | 8.82 | 8.76 | 8.151 | 7.851 | 7.496 | 6.669 | 6.1267 |
| 1500 | 8.4765 | 9.042 | 9.23 | 9.165 | 8.554 | 8.254 | 7.9 | 7.076 | 6.538 |
| 2000 | 8.7619 | 9.329 | 9.512 | 9.45 | 8.839 | 8.539 | 8.185 | 7.366 | 6.828 |
| 3000 | 9.1658 | 9.73 | 9.914 | 9.85 | 9.24 | 8.94 | 8.585 | 7.765 | 7.2411 |
| 4000 | 9.45 | 10.015 | 10.2 | 10.13 | 9.525 | 9.224 | 8.87 | 8.048 | 7.521 |
| 6000 | 9.852 | 10.418 | 10.6 | 10.54 | 9.928 | 9.627 | 9.271 | 8.45 | 7.919 |
| 8000 | 10.138 | 10.704 | 10.89 | 10.83 | 10.21 | 9.913 | 9.556 | 8.735 | 8.202 |

## Table 2.7 Vertical fracture. Drawdown in a monitoring well on y-axis.
Kruseman and de Ridder (2001).

| $t_D/r'$ | r'=0.05 | 0.07 | 0.1 | 0.2 | 0.3 | 0.5 | 1 | 2 | 5 |
|---|---|---|---|---|---|---|---|---|---|
| 0.001 | | | | | | | | | |
| 0.0015 | | | | | | | | | |
| 0.002 | | | | | | | | | |
| 0.003 | 0.00006 | 0.00001 | | | | | | | |
| 0.004 | 0.00025 | 0.00007 | 0.00001 | | | | | | |
| 0.006 | 0.00117 | 0.00047 | 0.00012 | | | | | | |
| 0.008 | 0.00275 | 0.00139 | 0.0005 | 0.00002 | | | | | |
| 0.01 | 0.00483 | 0.0028 | 0.00124 | 0.00008 | 0.00001 | | | | |
| 0.015 | 0.01129 | 0.00792 | 0.0046 | 0.00075 | 0.00013 | | | | |
| 0.02 | 0.0186 | 0.01442 | 0.00965 | 0.00248 | 0.00064 | 0.00004 | | | |
| 0.03 | 0.03376 | 0.02906 | 0.02259 | 0.00921 | 0.00372 | 0.00062 | 0.00001 | | |
| 0.04 | 0.04857 | 0.04423 | 0.03721 | 0.01931 | 0.00978 | 0.00252 | 0.000029 | | |
| 0.06 | 0.07619 | 0.07375 | 0.06751 | 0.04517 | 0.0289 | 0.01169 | 0.00123 | 0.00001 | |
| 0.08 | 0.101127 | 0.10137 | 0.09713 | 0.0744 | 0.054 | 0.0275 | 0.005 | 0.00017 | |
| 0.1 | 0.1243 | 0.1271 | 0.12544 | 0.1047 | 0.0821 | 0.04825 | 0.01221 | 0.00077 | |
| 0.15 | 0.1752 | 0.1849 | 0.1904 | 0.1797 | 0.157 | 0.11257 | 0.04425 | 0.00636 | 0.00002 |
| 0.2 | 0.2193 | 0.2357 | 0.2486 | 0.2509 | 0.2324 | 0.1844 | 0.0901 | 0.01954 | 0.00025 |
| 0.3 | 0.2951 | 0.3234 | 0.3503 | 0.3806 | 0.3747 | 0.329 | 0.199 | 0.0653 | 0.0029 |
| 0.4 | 0.3598 | 0.399 | 0.4387 | 0.496 | 0.503 | 0.4637 | 0.3113 | 0.126 | 0.0106 |
| 0.6 | 0.4697 | 0.5276 | 0.5898 | 0.6943 | 0.7242 | 0.698 | 0.5187 | 0.2607 | 0.042 |
| 0.8 | 0.56293 | 0.6371 | 0.7184 | 0.8607 | 0.9075 | 0.89 | 0.6974 | 0.3925 | 0.08854 |
| 1 | 0.6454 | 0.7337 | 0.8314 | 1.0039 | 1.0634 | 1.053 | 0.8516 | 0.5143 | 0.1428 |
| 1.5 | 0.8209 | 0.938 | 1.067 | 1.2928 | 1.3731 | 1.374 | 1.161 | 0.7746 | 0.287 |
| 2 | 0.96745 | 1.1063 | 1.2575 | 1.517 | 1.61 | 1.618 | 1.398 | 0.9849 | 0.4254 |
| 3 | 1.207 | 1.3757 | 1.555 | 1.856 | 1.9626 | 1.979 | 1.752 | 1.3096 | 0.6668 |
| 4 | 1.4 | 1.5877 | 1.785 | 2.109 | 2.223 | 2.245 | 2.013 | 1.555 | 0.8662 |
| 6 | 1.7 | 1.911 | 2.129 | 2.478 | 2.6 | 2.625 | 2.392 | 1.918 | 1.179 |
| 8 | 1.93 | 2.156 | 2.3843 | 2.7475 | 2.874 | 2.901 | 2.666 | 2.184 | 1.418 |
| 10 | 2.118 | 2.352 | 2.588 | 2.959 | 3.0888 | 3.117 | 2.8809 | 2.394 | 1.612 |
| 15 | 2.473 | 2.721 | 2.966 | 3.35 | 3.483 | 3.513 | 3.275 | 2.782 | 1.978 |
| 20 | 2.735 | 2.99 | 3.24 | 3.63 | 3.765 | 3.796 | 3.558 | 3.061 | 2.245 |
| 30 | 3.114 | 3.376 | 3.632 | 4.028 | 4.164 | 4.197 | 3.958 | 3.458 | 2.623 |
| 40 | 3.389 | 3.654 | 3.9125 | 4.3116 | 4.45 | 4.482 | 4.242 | 3.7408 | 2.907 |
| 60 | 3.781 | 4.049 | 4.311 | 4.713 | 4.852 | 4.885 | 4.645 | 4.142 | 3.302 |
| 80 | 4.062 | 4.332 | 4.595 | 5 | 5.139 | 5.174 | 4.9315 | 4.427 | 3.585 |
| 100 | 4.281 | 4.552 | 4.816 | 5.221 | 5.36 | 5.394 | 5.154 | 4.649 | 3.805 |
| 150 | 4.681 | 4.954 | 5.216 | 5.625 | 5.764 | 5.798 | 5.558 | 5.023 | 4.206 |
| 200 | 4.966 | 5.239 | 5.505 | 5.912 | 6.052 | 6.086 | 5.845 | 5.3396 | 4.492 |
| 300 | 5.368 | 5.643 | 5.909 | 6.316 | 6.456 | 6.491 | 6.25 | 5.744 | 4.896 |

**Table 2.7 continued.**

| $t_D/r'$ | r'=0.05 | 0.07 | 0.1 | 0.2 | 0.3 | 0.5 | 1 | 2 | 5 |
|---|---|---|---|---|---|---|---|---|---|
| 400 | 5.655 | 5.929 | 6.196 | 6.604 | 6.744 | 6.778 | 6.538 | 6.032 | 5.182 |
| 600 | 6.059 | 6.33 | 6.6 | 7.009 | 7.149 | 7.183 | 6.943 | 6.437 | 5.587 |
| 800 | 6.3456 | 6.62 | 6.888 | 7.2964 | 7.437 | 7.471 | 7.23 | 6.724 | 5.874 |
| 1000 | 6.568 | 6.843 | 7.111 | 7.519 | 7.659 | 7.694 | 7.449 | 6.948 | 6.097 |
| 1500 | 6.973 | 7.249 | 7.516 | 7.925 | 8.065 | 8.093 | 7.847 | 7.339 | 6.502 |
| 2000 | 7.2607 | 7.536 | 7.804 | 8.212 | 8.348 | 8.376 | 8.13 | 7.619 | 6.79 |
| 3000 | 7.6659 | 7.94 | 8.209 | 8.612 | 8.748 | 8.777 | 8.53 | 8.015 | 7.167 |
| 4000 | 7.9534 | 8.23 | 8.497 | 8.896 | 9.032 | 9.062 | 8.814 | 8.297 | 7.436 |
| 6000 | 8.359 | 8.634 | 8.896 | 9.297 | 9.434 | 9.464 | 9.217 | 8.696 | 7.819 |
| 8000 | 8.646 | 8.918 | 9.1805 | 9.583 | 9.719 | 9.75 | 9.502 | 8.98 | 8.095 |

## Table 2.8 Vertical fracture. Drawdown in a monitoring well on a 45° line.
Kruseman and de Ridder, (2001).

| t_D/r' | r'=0.05 | 0.07 | 0.1 | 0.2 | 0.3 | 0.5 | 1 | 2 | 5 |
|---|---|---|---|---|---|---|---|---|---|
| 0.002 | 0.00003 | | | | | | | | |
| 0.003 | 0.00023 | 0.00006 | 0.00001 | | | | | | |
| 0.004 | 0.00068 | 0.00026 | 0.00006 | | | | | | |
| 0.006 | 0.00227 | 0.0012 | 0.00046 | 0.00002 | | | | | |
| 0.008 | 0.00452 | 0.0028 | 0.00136 | 0.00013 | 0.00001 | | | | |
| 0.01 | 0.00714 | 0.00489 | 0.00275 | 0.0004 | 0.00006 | | | | |
| 0.015 | 0.01445 | 0.01139 | 0.00782 | 0.00217 | 0.00061 | 0.00005 | | | |
| 0.02 | 0.022 | 0.01872 | 0.01428 | 0.0055 | 0.0021 | 0.00031 | | | |
| 0.03 | 0.0368 | 0.03387 | 0.02889 | 0.01564 | 0.00825 | 0.00229 | 0.00009 | | |
| 0.04 | 0.0507 | 0.04866 | 0.0441 | 0.02856 | 0.01779 | 0.00679 | 0.00057 | | |
| 0.06 | 0.0757 | 0.07621 | 0.0736 | 0.05778 | 0.04284 | 0.02273 | 0.00397 | 0.00003 | |
| 0.08 | 0.0979 | 0.10121 | 0.1013 | 0.08814 | 0.0717 | 0.04511 | 0.01138 | 0.00022 | |
| 0.1 | 0.1181 | 0.1241 | 0.1271 | 0.1181 | 0.1018 | 0.07122 | 0.02242 | 0.00079 | |
| 0.15 | 0.1622 | 0.1748 | 0.1851 | 0.1892 | 0.1771 | 0.1432 | 0.0608 | 0.00513 | |
| 0.2 | 0.2 | 0.2188 | 0.2361 | 0.2543 | 0.2491 | 0.2165 | 0.1074 | 0.01438 | 0.00004 |
| 0.3 | 0.2647 | 0.2942 | 0.3243 | 0.3702 | 0.3802 | 0.3551 | 0.2078 | 0.04573 | 0.00063 |
| 0.4 | 0.3198 | 0.3587 | 0.4 | 0.4719 | 0.4964 | 0.4797 | 0.3071 | 0.08844 | 0.003 |
| 0.6 | 0.4128 | 0.468 | 0.5294 | 0.646 | 0.6948 | 0.6927 | 0.4891 | 0.1883 | 0.01599 |
| 0.8 | 0.4917 | 0.561 | 0.6393 | 0.7929 | 0.8603 | 0.8694 | 0.6478 | 0.2918 | 0.03984 |
| 1 | 0.56135 | 0.6428 | 0.7363 | 0.9206 | 1.002 | 1.02 | 0.7867 | 0.3914 | 0.07164 |
| 1.5 | 0.71 | 0.817 | 0.941 | 1.1824 | 1.289 | 1.32 | 1.07 | 0.6141 | 0.1686 |
| 2 | 0.835 | 0.963 | 1.11 | 1.39 | 1.511 | 1.551 | 1.293 | 0.8014 | 0.2722 |
| 3 | 1.042 | 1.2014 | 1.379 | 1.708 | 1.847 | 1.897 | 1.63 | 1.0997 | 0.4685 |
| 4 | 1.212 | 1.393 | 1.591 | 1.949 | 2.098 | 2.154 | 1.882 | 1.331 | 0.6405 |
| 6 | 1.484 | 1.692 | 1.915 | 2.306 | 2.465 | 2.528 | 2.25 | 1.678 | 0.9218 |
| 8 | 1.697 | 1.922 | 2.159 | 2.568 | 2.734 | 2.799 | 2.519 | 1.936 | 1.144 |
| 10 | 1.873 | 2.109 | 2.355 | 2.776 | 2.945 | 3.012 | 2.7311 | 2.141 | 1.327 |
| 15 | 2.212 | 2.464 | 2.724 | 3.16 | 3.334 | 3.404 | 3.121 | 2.521 | 1.677 |
| 20 | 2.464 | 2.726 | 2.993 | 3.437 | 3.614 | 3.685 | 3.401 | 2.797 | 1.936 |
| 30 | 2.833 | 3.105 | 3.379 | 3.832 | 4.011 | 4.084 | 3.8 | 3.19 | 2.313 |
| 40 | 3.103 | 3.379 | 3.657 | 4.114 | 4.295 | 4.368 | 4.083 | 3.471 | 2.586 |
| 60 | 3.489 | 3.7703 | 4.052 | 4.515 | 4.696 | 4.77 | 4.485 | 3.87 | 2.977 |
| 80 | 3.767 | 4.051 | 4.335 | 4.8 | 4.982 | 5.056 | 4.77 | 4.155 | 3.257 |
| 100 | 3.985 | 4.27 | 4.555 | 5.021 | 5.204 | 5.279 | 4.992 | 4.376 | 3.476 |
| 150 | 4.382 | 4.67 | 4.957 | 5.424 | 5.608 | 5.683 | 5.396 | 4.779 | 3.875 |
| 200 | 4.666 | 4.955 | 5.242 | 5.711 | 5.895 | 5.97 | 5.683 | 5.065 | 4.16 |
| 300 | 5.068 | 5.357 | 5.646 | 6.115 | 6.3 | 6.375 | 6.088 | 5.47 | 4.563 |

**Table 2.8 continued.**

| $t_D/r'$ | r'=0.05 | 0.07 | 0.1 | 0.2 | 0.3 | 0.5 | 1 | 2 | 5 |
|---|---|---|---|---|---|---|---|---|---|
| 400 | 5.353 | 5.644 | 5.932 | 6.403 | 6.587 | 6.662 | 6.374 | 5.756 | 4.849 |
| 600 | 5.757 | 6.048 | 6.337 | 6.808 | 6.992 | 7.067 | 6.777 | 6.161 | 5.253 |
| 800 | 6.044 | 6.335 | 6.624 | 7.095 | 7.279 | 7.353 | 7.064 | 6.452 | 5.54 |
| 1000 | 6.266 | 6.557 | 6.847 | 7.318 | 7.502 | 7.575 | 7.286 | 6.677 | 5.762 |
| 1500 | 6.671 | 6.963 | 7.252 | 7.723 | 7.906 | 7.98 | 7.69 | 7.086 | 6.167 |
| 2000 | 6.958 | 7.25 | 7.54 | 8.011 | 8.193 | 8.265 | 7.976 | 7.375 | 6.455 |
| 3000 | 7.363 | 7.655 | 7.945 | 8.414 | 8.596 | 8.666 | 8.375 | 7.777 | 6.871 |
| 4000 | 7.651 | 7.943 | 8.233 | 8.7 | 8.88 | 8.951 | 8.66 | 8.061 | 7.164 |
| 6000 | 8.056 | 8.348 | 8.637 | 9.1 | 9.281 | 9.352 | 9.06 | 8.462 | 7.561 |
| 8000 | 8.343 | 8.635 | 8.921 | 9.384 | 9.566 | 9.638 | 9.345 | 8.747 | 7.846 |

## Table 2.9 Vertical fracture with finite hydraulic conductivity.
Cinco, Samaniego and Dominguez (1978).

| t_D | 0.1π | 0.2π | 0.5π | π | 2π | 10π | 20π | 100π |
|---|---|---|---|---|---|---|---|---|
| 0.001 | 1.54 | 1.0898 | 0.66 | 0.4886 | 0.3464 | 0.1732 | 0.1436 | 0.118 |
| 0.002 | 1.8 | 1.276 | 0.78 | 0.5762 | 0.4112 | 0.22 | 0.1892 | 0.1628 |
| 0.003 | 1.94 | 1.4048 | 0.872 | 0.638 | 0.4578 | 0.2554 | 0.224 | 0.1972 |
| 0.004 | 2.04 | 1.504 | 0.932 | 0.6864 | 0.495 | 0.2848 | 0.253 | 0.226 |
| 0.005 | 2.156 | 1.5852 | 0.99 | 0.7266 | 0.5264 | 0.3106 | 0.2784 | 0.2512 |
| 0.006 | 2.238 | 1.6546 | 1.028 | 0.7612 | 0.554 | 0.3336 | 0.3012 | 0.2738 |
| 0.007 | 2.31 | 1.7152 | 1.072 | 0.7918 | 0.5786 | 0.3546 | 0.322 | 0.2944 |
| 0.008 | 2.376 | 1.7692 | 1.11 | 0.8196 | 0.6012 | 0.3742 | 0.3412 | 0.3136 |
| 0.009 | 2.44 | 1.818 | 1.14 | 0.8448 | 0.622 | 0.3924 | 0.3592 | 0.3316 |
| 0.01 | 2.5 | 1.8626 | 1.17 | 0.8682 | 0.6414 | 0.4094 | 0.3762 | 0.3484 |
| 0.02 | 2.92 | 2.1674 | 1.404 | 1.0362 | 0.7878 | 0.544 | 0.5098 | 0.4814 |
| 0.03 | 3.18 | 2.3722 | 1.562 | 1.1576 | 0.896 | 0.6442 | 0.6094 | 0.5806 |
| 0.04 | 3.4 | 2.5276 | 1.684 | 1.2544 | 0.984 | 0.7266 | 0.6912 | 0.662 |
| 0.05 | 3.55 | 2.6538 | 1.8 | 1.3364 | 1.0594 | 0.7976 | 0.7618 | 0.7324 |
| 0.06 | 3.66 | 2.7804 | 1.88 | 1.408 | 1.126 | 0.8608 | 0.8244 | 0.7948 |
| 0.07 | 3.76 | 2.8532 | 1.96 | 1.4722 | 1.1858 | 0.9176 | 0.8812 | 0.851 |
| 0.08 | 3.87 | 2.9356 | 2.02 | 1.5306 | 1.2406 | 0.9698 | 0.933 | 0.9028 |
| 0.09 | 3.96 | 3.0098 | 2.08 | 1.5842 | 1.2912 | 1.018 | 0.981 | 0.9506 |
| 0.1 | 4.036 | 3.0774 | 2.14 | 1.634 | 1.3382 | 1.0632 | 1.0258 | 0.995 |
| 0.2 | 4.56 | 3.5456 | 2.54 | 2.002 | 1.6906 | 1.403 | 1.364 | 1.3322 |
| 0.3 | 4.9 | 3.8526 | 2.8 | 2.2578 | 1.9372 | 1.6416 | 1.6016 | 1.569 |
| 0.4 | 5.14 | 4.0828 | 3.02 | 2.4564 | 2.1296 | 1.8286 | 1.788 | 1.7548 |
| 0.5 | 5.32 | 4.268 | 3.2 | 2.6198 | 2.2884 | 1.9836 | 1.9424 | 1.9088 |
| 0.6 | 5.5 | 4.423 | 3.35 | 2.7588 | 2.4242 | 2.1164 | 2.0748 | 2.041 |
| 0.7 | 5.64 | 4.5568 | 3.46 | 2.8802 | 2.543 | 2.2326 | 2.191 | 2.1568 |
| 0.8 | 5.8 | 4.6744 | 3.56 | 2.9878 | 2.6486 | 2.3364 | 2.2944 | 2.26 |
| 0.9 | 5.9 | 4.7794 | 3.66 | 3.0848 | 2.7438 | 2.43 | 2.3878 | 2.3532 |
| 1 | 6 | 4.8742 | 3.76 | 3.173 | 2.8306 | 2.5154 | 2.473 | 2.4384 |
| 2 | 6.7 | 5.5158 | 4.36 | 3.7812 | 3.4312 | 3.1092 | 3.0658 | 3.0304 |
| 3 | 7.1 | 5.904 | 4.76 | 4.1566 | 3.8036 | 3.4788 | 3.435 | 3.3992 |
| 4 | 7.4 | 6.1828 | 5.06 | 4.4288 | 4.0742 | 3.7478 | 3.7038 | 3.668 |
| 5 | 7.6 | 6.4004 | 5.3 | 4.6424 | 4.287 | 3.9596 | 3.9154 | 3.8796 |
| 6 | 7.78 | 6.5792 | 5.49 | 4.8184 | 4.4622 | 4.1342 | 4.09 | 4.054 |
| 7 | 7.96 | 6.7308 | 5.6 | 4.9678 | 4.6112 | 4.2828 | 4.2386 | 4.2026 |

## Table 2.9 continued.

| $t_D$ | $0.1\pi$ | $0.2\pi$ | $0.5\pi$ | $\pi$ | $2\pi$ | $10\pi$ | $20\pi$ | $100\pi$ |
|---|---|---|---|---|---|---|---|---|
| 8 | 8.1 | 6.8624 | 5.74 | 5.098 | 4.741 | 4.4122 | 4.3678 | 4.3318 |
| 9 | 8.2 | 6.9782 | 5.84 | 5.213 | 4.8558 | 4.5268 | 4.4824 | 4.4662 |
| 10 | 8.3 | 7.0828 | 5.96 | 5.3162 | 4.9588 | 4.6294 | 4.585 | 4.549 |
| 20 | 9 | 7.7704 | 6.7 | 5.9996 | 5.641 | 5.3106 | 5.2662 | 5.23 |
| 30 | 9.4 | 8.174 | 7.08 | 6.4016 | 6.0428 | 5.7122 | 5.6676 | 5.6314 |
| 40 | 9.72 | 8.4608 | 7.32 | 6.6878 | 6.3286 | 5.9978 | 5.9532 | 5.917 |
| 50 | 9.98 | 8.6834 | 7.55 | 6.9098 | 6.5506 | 6.2198 | 6.1752 | 6.139 |
| 60 | 10.18 | 8.8654 | 7.78 | 7.0936 | 6.7322 | 6.4014 | 6.3568 | 6.3204 |
| 70 | 10.34 | 9.0194 | 7.9 | 7.2452 | 6.886 | 6.555 | 6.5102 | 6.474 |
| 80 | 10.44 | 9.1526 | 8.04 | 7.3784 | 7.019 | 6.688 | 6.6434 | 6.607 |
| 90 | 10.58 | 9.2702 | 8.16 | 7.496 | 7.1366 | 6.8054 | 6.7608 | 6.7246 |
| 100 | 10.7 | 9.3756 | 8.24 | 7.601 | 7.2416 | 6.9106 | 6.866 | 6.8296 |
| 200 | 11.46 | 10.0682 | 8.94 | 8.2932 | 7.9336 | 7.6026 | 7.5578 | 7.5216 |
| 300 | 11.9 | 10.4734 | 9.4 | 8.6984 | 8.3388 | 8.0076 | 7.963 | 7.9266 |
| 400 | 12.12 | 10.761 | 9.64 | 8.9858 | 8.6262 | 8.295 | 8.2504 | 8.214 |
| 500 | 12.3 | 10.984 | 9.92 | 9.209 | 8.8494 | 8.518 | 8.4734 | 8.4372 |
| 600 | 12.46 | 11.1664 | 10.04 | 9.3912 | 9.0316 | 8.7004 | 8.6556 | 8.6194 |
| 700 | 12.6 | 11.3204 | 10.2 | 9.5452 | 9.1856 | 8.8544 | 8.8098 | 8.7734 |
| 800 | 12.7 | 11.454 | 10.34 | 9.6788 | 9.3192 | 8.9878 | 8.9432 | 8.907 |
| 900 | 12.8 | 11.5718 | 10.4 | 9.7966 | 9.437 | 9.1056 | 9.061 | 9.0246 |
| 1000 | 12.96 | 11.6766 | 10.5 | 9.9018 | 9.5422 | 9.211 | 9.1662 | 9.13 |

## Table 2.10 Drawdown in a well in a horizontal fracture at center of formation. Gringarten and Ramey (1974).

| t_D | h_D = 0.05 | 0.1 | 0.2 | 0.3 | 0.4 | 0.5 | 0.7 | 1 |
|---|---|---|---|---|---|---|---|---|
| 0.001 | 0.099 | 0.074 | 0.069 | 0.069 | 0.069 | 0.069 | 0.069 | 0.069 |
| 0.002 | 0.176 | 0.116 | 0.099 | 0.099 | 0.099 | 0.099 | 0.099 | 0.099 |
| 0.003 | 0.255 | 0.157 | 0.123 | 0.123 | 0.123 | 0.123 | 0.123 | 0.123 |
| 0.004 | 0.332 | 0.193 | 0.143 | 0.143 | 0.143 | 0.143 | 0.143 | 0.143 |
| 0.006 | 0.486 | 0.272 | 0.186 | 0.175 | 0.175 | 0.175 | 0.175 | 0.175 |
| 0.008 | 0.642 | 0.35 | 0.225 | 0.205 | 0.203 | 0.203 | 0.203 | 0.203 |
| 0.01 | 0.816 | 0.425 | 0.267 | 0.232667 | 0.2265 | 0.2256 | 0.225714 | 0.2258 |
| 0.02 | 1.616 | 0.82 | 0.467 | 0.366667 | 0.333 | 0.3224 | 0.319429 | 0.3192 |
| 0.03 | 2.416 | 1.24 | 0.667 | 0.5 | 0.4335 | 0.406 | 0.392286 | 0.391 |
| 0.04 | 3.216 | 1.66 | 0.867 | 0.633333 | 0.5335 | 0.4864 | 0.456286 | 0.4514 |
| 0.06 | 4.816 | 2.42 | 1.267 | 0.9 | 0.7335 | 0.6468 | 0.574286 | 0.5536 |
| 0.08 | 6.356 | 3.2 | 1.652 | 1.156667 | 0.926 | 0.8008 | 0.688571 | 0.6414 |
| 0.1 | 7.856 | 3.95 | 2.027 | 1.406667 | 1.1135 | 0.9508 | 0.793429 | 0.7208 |
| 0.2 | 14.36 | 7.2 | 3.653 | 2.490667 | 1.9265 | 1.6012 | 1.258 | 1.0494 |
| 0.3 | 19.436 | 9.65 | 4.922 | 3.336667 | 2.561 | 2.1088 | 1.620571 | 1.3032 |
| 0.4 | 23.532 | 11.8 | 5.946 | 4.019333 | 3.073 | 2.5184 | 1.913143 | 1.508 |
| 0.6 | 29.88 | 15.2 | 7.533 | 5.077333 | 3.8665 | 3.1532 | 2.366571 | 1.8254 |
| 0.8 | 34.708 | 17.8 | 8.74 | 5.882 | 4.47 | 3.636 | 2.711429 | 2.0668 |
| 1 | 38.6 | 20 | 9.713 | 6.530667 | 4.9565 | 4.0252 | 2.989429 | 2.2614 |
| 2 | 51.304 | 26.3 | 12.889 | 8.648 | 6.5445 | 5.2956 | 3.896857 | 2.8966 |
| 3 | 59.004 | 30 | 14.814 | 9.931333 | 7.507 | 6.0656 | 4.446857 | 3.2816 |
| 4 | 64.956 | 32.6 | 16.302 | 10.92333 | 8.251 | 6.6608 | 4.872 | 3.5792 |
| 6 | 72.46 | 36.6 | 18.178 | 12.174 | 9.189 | 7.4112 | 5.408 | 3.9544 |
| 8 | 78.112 | 39.3 | 19.591 | 13.116 | 9.8955 | 7.9764 | 5.811714 | 4.237 |
| 10 | 82.512 | 42 | 20.691 | 13.84933 | 10.4455 | 8.4164 | 6.126 | 4.457 |
| 20 | 96.264 | 48.3 | 24.129 | 16.14133 | 12.1645 | 9.7916 | 7.108286 | 5.1446 |
| 30 | 104.336 | 52.5 | 26.147 | 17.48667 | 13.1735 | 10.5988 | 7.684857 | 5.5482 |
| 40 | 110.068 | 55.8 | 27.58 | 18.442 | 13.89 | 11.172 | 8.094286 | 5.8348 |
| 60 | 118.156 | 60 | 29.602 | 19.79 | 14.901 | 11.9808 | 8.672 | 6.2392 |
| 80 | 123.9 | 63 | 31.038 | 20.74733 | 15.619 | 12.5552 | 9.082286 | 6.5264 |
| 100 | 128.356 | 66 | 32.152 | 21.49 | 16.176 | 13.0008 | 9.400571 | 6.7492 |
| 200 | 142.22 | 74 | 35.618 | 23.80067 | 17.909 | 14.3872 | 10.39086 | 7.4424 |
| 300 | 150.328 | 79 | 37.645 | 25.152 | 18.9225 | 15.198 | 10.97 | 7.8478 |
| 400 | 156.08 | 83 | 39.083 | 26.11067 | 19.6415 | 15.7732 | 11.38086 | 8.1354 |
| 600 | 164.188 | 88 | 41.11 | 27.462 | 20.655 | 16.584 | 11.96 | 8.5408 |
| 800 | 169.94 | 91 | 42.548 | 28.42067 | 21.374 | 17.1592 | 12.37086 | 8.8284 |
| 1000 | 174.4 | 93 | 43.663 | 29.164 | 21.9315 | 17.6052 | 12.68943 | 9.0514 |

**Table 2.10 continued.**

| $t_D$ | $h_D =$ 1.5 | 2 | 3 | 4 | 5 | 6 | 7 | 10 |
|---|---|---|---|---|---|---|---|---|
| 0.001 | 0.069 | 0.069 | 0.069 | 0.069 | 0.069 | 0.069 | 0.069 | 0.0706 |
| 0.002 | 0.099 | 0.099 | 0.099 | 0.099 | 0.099 | 0.099 | 0.099 | 0.099 |
| 0.003 | 0.123 | 0.123 | 0.123 | 0.123 | 0.123 | 0.123 | 0.123 | 0.123 |
| 0.004 | 0.143 | 0.143 | 0.143 | 0.143 | 0.143 | 0.143 | 0.143 | 0.143 |
| 0.006 | 0.175 | 0.175 | 0.175 | 0.175 | 0.175 | 0.175 | 0.175 | 0.175 |
| 0.008 | 0.203 | 0.203 | 0.203 | 0.203 | 0.203 | 0.203 | 0.203 | 0.203 |
| 0.01 | 0.2257 | 0.2257 | 0.225733 | 0.2257 | 0.22572 | 0.2257 | 0.225714 | 0.22574 |
| 0.02 | 0.3192 | 0.3192 | 0.3192 | 0.3192 | 0.3192 | 0.3192 | 0.3192 | 0.31926 |
| 0.03 | 0.3909 | 0.3909 | 0.390933 | 0.39095 | 0.39096 | 0.390933 | 0.390943 | 0.38698 |
| 0.04 | 0.4514 | 0.4514 | 0.451333 | 0.45135 | 0.45136 | 0.451367 | 0.451371 | 0.4514 |
| 0.06 | 0.5521 | 0.5521 | 0.552067 | 0.5507 | 0.55208 | 0.5514 | 0.552086 | 0.5521 |
| 0.08 | 0.6352 | 0.6352 | 0.6352 | 0.6352 | 0.63516 | 0.635167 | 0.635171 | 0.6352 |
| 0.1 | 0.7058 | 0.7058 | 0.7058 | 0.7056 | 0.70584 | 0.705833 | 0.705829 | 0.70586 |
| 0.2 | 0.95 | 0.9488 | 0.948 | 0.948 | 0.948 | 0.947967 | 0.947971 | 0.94802 |
| 0.3 | 1.12 | 1.0986 | 1.092533 | 1.09255 | 1.09252 | 1.0925 | 1.092514 | 1.09256 |
| 0.4 | 1.26 | 1.2079 | 1.191067 | 1.1907 | 1.19068 | 1.190667 | 1.190829 | 1.1907 |
| 0.6 | 1.46 | 1.3693 | 1.321733 | 1.31865 | 1.31856 | 1.318567 | 1.318543 | 1.31852 |
| 0.8 | 1.63 | 1.4903 | 1.410067 | 1.40105 | 1.4004 | 1.4004 | 1.400543 | 1.40042 |
| 1 | 1.76 | 1.5877 | 1.4776 | 1.4605 | 1.4586 | 1.458467 | 1.458971 | 1.45848 |
| 2 | 2.18 | 1.9053 | 1.6908 | 1.6305 | 1.61424 | 1.6105 | 1.609771 | 1.6097 |
| 3 | 2.45 | 2.0978 | 1.819133 | 1.7274 | 1.6948 | 1.683767 | 1.6804 | 1.67928 |
| 4 | 2.64 | 2.2466 | 1.918333 | 1.80185 | 1.75084 | 1.732167 | 1.725 | 1.72146 |
| 6 | 2.95 | 2.4342 | 2.0434 | 1.89565 | 1.82996 | 1.798767 | 1.783829 | 1.7706 |
| 8 | 3.15 | 2.5755 | 2.1376 | 1.9663 | 1.88648 | 1.8459 | 1.824514 | 1.80458 |
| 10 | 3.32 | 2.6855 | 2.210933 | 2.0213 | 1.93048 | 1.8826 | 1.856 | 1.8279 |
| 20 | 3.8 | 3.0293 | 2.440133 | 2.1932 | 2.068 | 1.9972 | 1.954257 | 1.89754 |
| 30 | 4.05 | 3.2311 | 2.574667 | 2.2941 | 2.14872 | 2.064467 | 2.011914 | 1.9379 |
| 40 | 4.25 | 3.3744 | 2.6702 | 2.36575 | 2.20604 | 2.112233 | 2.052857 | 1.96656 |
| 60 | 4.52 | 3.5766 | 2.805 | 2.46685 | 2.28692 | 2.179633 | 2.110629 | 2.007 |
| 80 | 4.7 | 3.7202 | 2.900733 | 2.53865 | 2.34436 | 2.2275 | 2.151657 | 2.03572 |
| 100 | 4.88 | 3.8316 | 2.975 | 2.59435 | 2.38892 | 2.264633 | 2.183486 | 2.058 |
| 200 | 5.3 | 4.1782 | 3.206067 | 2.76765 | 2.52756 | 2.380167 | 2.282514 | 2.12732 |
| 300 | 5.61 | 4.3809 | 3.3412 | 2.869 | 2.60864 | 2.447733 | 2.340429 | 2.16786 |
| 400 | 5.8 | 4.5247 | 3.437067 | 2.9409 | 2.66616 | 2.495667 | 2.381514 | 2.19662 |
| 600 | 6.15 | 4.7274 | 3.5722 | 3.04225 | 2.74724 | 2.563233 | 2.439429 | 2.23716 |
| 800 | 6.35 | 4.8712 | 3.668067 | 3.11415 | 2.80476 | 2.611167 | 2.480514 | 2.26592 |
| 1000 | 6.51 | 4.9827 | 3.7424 | 3.1699 | 2.84936 | 2.648333 | 2.512371 | 2.28822 |

ISBN 142513019-4
9 781425 130190

Printed in Great Britain
by Amazon.co.uk, Ltd.,
Marston Gate.